Revision Notes
on
Building Measurement

Trevor J. Saunt FCIOB, MBIM

Principal Lecturer, Department of Building and Environmental Health, Trent Polytechnic, Nottingham

BUTTERWORTHS
LONDON   BOSTON
Sydney   Wellington   Durban   Toronto

First published 1981

© Butterworth & Co. (Publishers) Ltd, 1981

British Library Cataloguing in Publication Data

Saunt, Trevor James
    Revision notes on building measurement.
    1. Building – Estimates – Great Britain
    I. Title
    690'.028'7   TH435   80-41296

    ISBN 0-408-00277-8

Typeset by Butterworths Litho Preparation Department
Printed in England by Fakenham Press Ltd., Fakenham, Norfolk

Preface

This book outlines the main aspects of the measurement of building work as studied by building technicians following the guide syllabuses laid down by the Technician Education Council, for the subject of 'Measurement' in the courses leading to the award of Ordinary or Higher Certificates or Diplomas in Building. In addition, it provides a basis of study for students studying the same subject leading to the Institute of Building Licentiate and Final Part I professional examinations.

Although the book includes a series of schedules that summarise the requirements of certain aspects of the *Standard Method of Measurement of Building Work*, it should not be used as a substitute for this essential reference document, but rather should be read in conjunction with it.

The process of the measurement of builders' work has been in operation for many years, and it should be appreciated that there are several alternative ways of setting down the dimensions and descriptions. One of the objectives of this book is not to set out a specific method as a rigid system, but rather to try to develop in the student a critical and systematic approach to the measurement procedure in order that the work may be carried out as effectively and as accurately as possible.

I am indebted to many of my past and present colleagues for their useful comments and criticisms during the preparation of this manuscript. In particular I am most grateful to Mr I. J. Wright for his work in preparing the drawings. I would also like to express my thanks to the Standing Joint Committee for the Standard Method of Measurement of Building Works for their permission to refer in the text of this book to the Sixth Edition of the *Standard Method of Measurement of Building Works*.

T. J. S.

Contents

1 Introduction to measurement

Bill of quantities

Basically a bill of quantities is a list or schedule of items of work comprising the extent of the contract. It should cover all the factors that will affect the cost of the building work: *labour* (both direct and indirect), *materials* and *plant* costs.

In compiling the total cost of the works there will be certain items for which it will be difficult either (a) to put an accurate price to the item, or (b) to charge its full cost against any particular trade. As a result the bill of quantities usually contains two sections:

1. The *preliminaries bill* includes the items mentioned in (b).
2. The *measured bill*, in addition to the measured items (which usually constitute the bulk of the contract), also includes the items mentioned in (a).

In some circumstances a separate section may be included for large volumes of work that are to be carried out on a daywork basis.

The bill of quantities constitutes one of the legal contract documents, and it is generally recommended that there should be no tendering in competition without this document, although no minimum contract value is applied to this recommendation as was the previous practice.

Types of bill of quantities

1. *Trade order or traditional bill.* This is set down in trade sequence, following the outline of the *Standard Method of Measurement*. It provides a good basis for competitive tendering, but leaves much to be desired as an aid to the general organisation of the contract and the control of finance.
2. *Elemental bill.* Again the work is billed in trade order – not in the sequence of the *Standard Method of Measurement*, but under headings that form a breakdown of the building into the main elements of the structure, i.e. walls, floors and roof. Although this provides a more useful control document during the construction operations, it makes the tendering process more complicated.
3. *Operational bill.* Work is divided into site operations or stages of the work, following the sequence of the construction programme. This is a list of the labour operations and materials involved, and where applicable plant requirements for each operation are set down separately. This facilitates a complete breakdown of costs, which then provides the most accurate method of controlling the organisation and financing of the contract.

In recent years the introduction of computers as an aid in the preparation of bills of quantities has led to the development of 'standard phraseology' or a 'library of standard descriptions', i.e. those developed in the document *Standard Phraseology for Bills of Quantities* by Fletcher and Moore.

Functions of the bill of quantities

The main function of the document is to provide a uniform basis by which all the contractors may tender competitively for the works.

As previously mentioned it provides a schedule of the items of work and may be used as a means of controlling the financing of the contract during the works, during preparation of valuations, pricing variations and implementing general cost control techniques. Upon completion of the works it may be used during the preparation of the Final Account, and afterwards possibly as a guide for future tendering. This last use basically depends upon the continual recording of accurate cost data during the construction programme.

Remember that the bill of quantities is a means of communication between the architect (working drawings) and the contractor's estimator (pricing of the works). If this communication process fails, the client may be faced with considerable additional costs of 'extras' or 'variations' that will be necessary during the constructional operations.

Reference documents

The main reference documents dealing with the measurement of building works are the *Standard Method of Measurement of Building Works*, and the *Code for the Measurement of Building Works in Small Dwellings*.

1. *Standard Method of Measurement*. Provides a uniform basis for measuring building works, laying down the recommended rules for measurement of all aspects of constructional work. By this means it ensures that the work is measured in a standard procedure with recognised units of measure and reasonably standard descriptions, thus enabling all contractors tendering for a contract to base their competitive tender figures on the same basic schedule of work. The recommendations of this document can be applied to either
    (a) the pre-measure of proposed constructional operations, or
    (b) the re-measure of work actually completed.
   All items of building work are measured as linear, superficial, cubic or enumerated items, and are then reduced for billing purposes into the basic units of either the metre or the kilogram.
2. *Code for the Measurement of Building Works in Small Dwellings*. Used in relation to the construction of small dwellings such as blocks of maisonettes or flats that do not exceed 130 m² floor area in each dwelling.

Role of the quantity surveyor

Covers a wide range of aspects throughout the complete cycle of a building project, namely:

1. *Initial cost advice* during the precontract stage – including preparation of approximate estimates, cost planning information concerned with the development of a design that provides the client with the best value for money.
2. *Tendering* – preparation of the necessary tender documents, advice on the various methods of obtaining tenders, critical examination of the tenders and priced bills of quantities when submitted.
3. *Valuation* of the work during the construction stage at regular intervals, and recommendation of the amounts to be paid to the main contractor. This stage is subsequently completed by the preparation of the Final Account at the end of the defects liability period.
4. *Cost control* – the quantity surveyor may also be called upon at any time during the construction stage to offer suggestions as to the economics of the adoption of alternative methods of construction or uses of materials.

Requirements of quantity surveying practice

The essential attributes of a good quantity surveyor when measuring the quantities of work are:

- A thorough knowledge of construction.
- The ability to scale off dimensions accurately from the drawings.

2

- The ability to put into words (accurate descriptions) what the architect has shown on the drawings.
- A consistent system of setting down the measurements and descriptions – i.e. logical sequence of measurement, setting down waste calculations, bracketing together items etc.
- A neat and methodical approach.

The main thing to remember is that other people will have to refer to these items several times upon subsequent occasions.

Preparation of bills of quantities

This involves two stages:

1. *Taking-off* – measurement of the work.
2. *Working-up*
   (a) *squaring* of dimensions, i.e. calculating lengths, areas and volumes of work set against each description and entering them in column D on the taking-off paper.
   (b) *Abstracting* – gathering together similar items to produce the total quantities for inclusion in the bill of quantities.
   (c) *Billing* – presentation or setting down of the items in the normal bill sequence.

All these processes require standard types of paper.

Taking-off

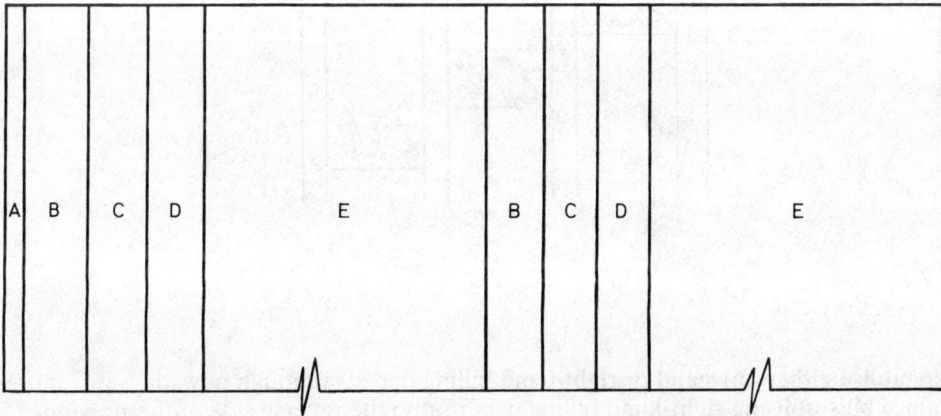

A Binding column.
B *Timesing column*, used as a means of multiplying a group of dimensions set against the description of an item of work that is known to be repeated several times within the contract. This eliminates repeatedly writing out descriptions unnecessarily.
C Used to record the sets of dimensions relating to the descriptions of items of work. These may be lineal, square or cubic, and dimensions may also be grouped.
D *Squaring column*, in which sets of dimensions are multiplied out and grouped dimensions are totalled to produce the overall quantities of the items of works.
E Used to write out the descriptions of the items of work; may also be used to set down (initial) waste calculations.

The taking-off sheet contains two sets of columns B to E. Bookings start at the top of the left-hand set of columns, continue down the sheet, and then follow on at the top of the right-hand set of columns.

To provide an easily recognisable referencing system during the later abstracting process, each set of columns (rather than each page) is given a number.

Abstracting

This involves the transfer of the quantities of similar items on to special paper, where the total quantities for entry into the bill of quantities are determined.

During the transfer process the column reference numbers from the taking-off sheets should also be noted, to make it easy to check the work later. In addition the items should also be clearly crossed out on the taking-off sheets.

Example

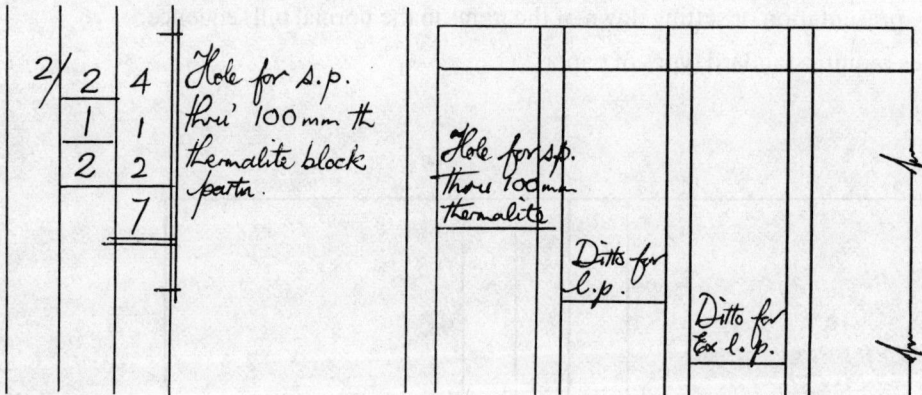

Billing

This process may utilise either left-hand or right-hand billing paper, although nowadays it is more common to produce bills utilising right-hand billing paper only: the reverse side of the previous pages can then be used to break down the unit rates into their labour, plant and material elemental costs. This facilitates control of costs during the construction programme.

At the end of the abstracting process, and prior to billing, the total quantities of the various items must be reduced to the unit of billing recommended in the appropriate section of the *Standard Method of Measurement*.

4

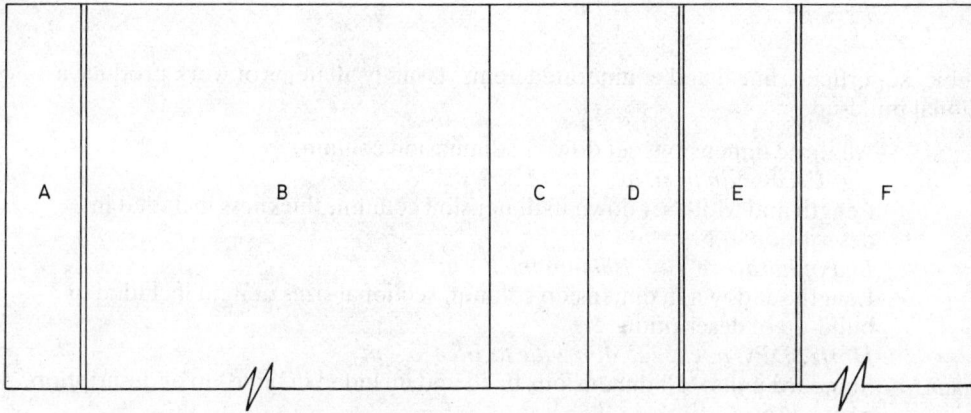

A Item reference, usually a letter; a combined page number and reference letter may then be allocated to each item.
B Description of the item of work.
C Total quantity of the item of work.
D Unit of measure, i.e. m^2, m^3 etc.
E Unit rate for the item of work.
F Total price for the item of work.

Principles of measurement of building works

Before studying the rules of measurement laid down under the various trade headings in the *Standard Method of Measurement*, students should be familiar with the 'General Rules' laid down in Section A of this document. These rules deal mainly with:

	SMM clause
1. *Bill of quantities* – must provide comprehensive description of the quality of work, together with an accurate representation of the actual quantities of work.	A2(1)
2. *Measurement* – work shall be taken to the nearest 10 mm and measured net (as fixed in position). Classification of dimensions and sizes, where given in groupings with a set lower and upper limit, shall be taken as exceeding the lower limit and not exceeding the upper limit.	A3(2)
3. *Descriptions* – recommended sequence of setting down dimensions in descriptions, i.e. length, width, height. All items shall be deemed to include for all the elements of production, i.e. labour, materials and plant, with an appropriate allowance for overheads and profit.	A3(4) A4(1)
	A2(2)
4. *Drawn information* – (a) types of drawings as classified in BS 1192 and used at the stage of preparation of the Bill of Quantities; (b) bill diagrams, i.e. use of sketched information in conjunction with the billed descriptions.	A5(1)
5. *Quantities* – rounding-off of quantities and the recognised units of billing.	A7(1, 2 & 3)
6. *Provisional and prime cost sums* – definitions of the types of work or services to be covered by such sums.	A8(1a & b)

5

Dimensions

Four types: cubic, superficial, lineal and enumerated items. Usually all items of work produce a three-dimensional build-up.

Cubic   All three dimensions set down in dimension column,
                           e.g. *Excav. fdn tr. n. ex. 1.00 m dp – m³*

Superficial   Length and width set down in dimension column, thickness included in
                           description, e.g.
                           Bed of hardcore – av. 100 mm thick – m²

Lineal   Length set down in dimension column, sectional sizes of item included in
                           build-up of description, e.g.
                           Horiz. DPC n. ex. 225 mm wide to BS 473 – m

Enumerated items   Standard units – all dimensions fixed and included in build-up of description,
                           e.g.
                           Clay chimney pot to BS 1181
                           300 mm long and 216 mm Ø – No.

Description

The general requirements are that the descriptions should be:

- *Concise* – but easily understood.
- *Specific* – should not allow any ambiguous interpretations.
- *Complete* – should eliminate the need for extras.
- *Accurate* – should state precisely the types and quality of materials etc. used.

Timesing

This process is used where the same item repeatedly occurs during the measurement of the work. It is used as a means of avoiding the monotonous repetition of the same descriptions, thus speeding up the measurement procedure.

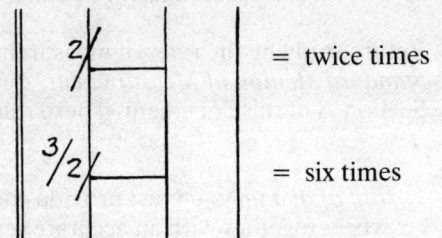

= twice times

= six times

The timesing column may also be used to 'dot on' similar items which it is found may be added to a previously written description of work:

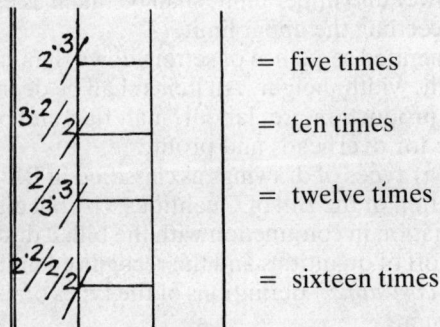

= five times

= ten times

= twelve times

= sixteen times

6

Bracketing together

This process is used as a means of maintaining clarity of presentation in order that it is perfectly clear which sets of dimensions relate to which descriptions. Generally bracketing is used:

- where more than one set of dimensions relate to one description, or
- where more than one description relates to the same set of dimensions.

Sometimes it may be necessary to amend dimensions where incorrect values have been entered on the take-off sheet. This should be done neatly and simply by cancelling the incorrect set of dimensions by writing alongside them the word 'NIL' and entering below the revised values. No attempt should be made to either change or erase incorrect values, as this may lead to misinterpretation of the information by other parties at some later stage of the work.

   Ideally dimensions booked down in the taking-off should be those shown on the drawings rather than scaled dimensions. Where it is necessary to calculate dimensions from groups of values shown on the drawings, these waste (preliminary) calculations should be entered to the right-hand side of the description column. This part of the sheet may also be used to enter other explanatory information.

Titles/headings

All taking-off sheets should be clearly headed or referenced to the contract to which they refer. In addition sub-headings should be used where possible to develop an easily recognisable or understandable sequence to the taking-off. This may be related to either the buildings comprising the contract or to recognised stages of the work such as those set out in the *Standard Method of Measurement*.

Deductions

From time to time it becomes necessary to make deductions for the previous over-measure of certain items of work. Ideally these deductions where possible should follow on immediately after the main (over-measured) item, although this is not always possible. For example, walls are initially measured as though no doors or windows existed; later, when doors and windows themselves are measured, the adjustment is made for the previous over-measure of brickwork.

ADJUSTMENT FOR OPENING

3/ 0.90
2.10

Ddt HB skin of holl wall in sand fcgs, in g.m (1:1:6) a.b.d.
&
Ddt Ditto in comms in g.m (1:1:6) a.b.d.
&
Ddt formation of 50 mm cav to holl wall inclg 4 N° wall ties to BS 1243

'Extra over' items

In measuring some items of work the *Standard Method of Measurement* allows for the description of the item as 'extra over'. This means that when pricing the work the value calculated is only the extra cost above that of the value set against the previously measured item. Examples are the measurement of:

- Fittings on rainwater pipes/gutters.
- Bends on pipework in plumbing installations.
- Extra over common brickwork for fairface and pointing.

Standard abbreviations

To reduce the work involved in the taking-off process, many abbreviations are used within the build-up of the descriptions. The only recognised document to include a list of abbreviations is BS 1192, but the following are some of the commonly adopted abbreviations now used:

a.b.	as before	br	bearer
a.b.d.	as before described	b.s.	both sides
agg.	aggregate	bwk	brickwork
ard	around	c. & f.	cut and fix
asp.	asphalt	c. & p.	cut and pin
av.	average	cap.	capillary
b.i.	built in	carcassg	carcassing
bit.	bitumen	cav.	cavity
bk	brick	c/cs	centres
bdg	boarding	C.C.N.	close copper nailing
bkt	bracket	c.i.	cast iron
bldg	building	chy	chimney
blk	blockwork	circ.	circular
b.m.s.	both measured separately	clg	ceiling
bottm	bottom	clse bdd	close boarded

c.m.	cement mortar	med.	medium
comms	commons	m.gd	make good
conc.	concrete	m.h.	manhole
cop.	copper	min.	minimum
c.p.	chromium plated	m.s.	measured separately
c.s.a.	cross-sectional area	nec.	necessary
cupd	cupboard	n.ex.	not exceeding
c.w.	cold water	nom.	nominal
c.w.b.g.	copper wire balloon grating	o/a	overall
ddt	deduct	O.C.N.	open copper nailing
ditto	as stated previously	o.e.	one end
dp	deep	o.e.	other end
DPC	damp proof course	opgs	openings
DPM	damp proof membrane	o.s.	one side
dr	door	o.s.	other side
ea.	each	p.c.	prime cost
earthwk	earthwork	perf.	perforation
e.g.	eaves gutter	p.f.	plain face
emulsn	emulsion	pla.	plaster
e.o.	extra over	p.m.	purpose made
ex.	exceeding	pr	pair
excav.	excavate	prov. s.	provisional sum
extg	existing	pt	paint
extl	external	ptg	pointing
fcgs	facings	ptn	partition
fdn	foundation	q.t.	quarry tile
f.f.	fair face	r.conc. or RC	reinforced concrete
fin.	finish	rec.	receive
flg	flooring	reinfm't	reinforcement
fmwk	formwork	r.o.j.	rake out joints
f.o.	fix only	r.w.p.	rainwater pipe
follg	following	s.c.	stop cock
fxd	fixed	s.e.	stopped end
galv'd	galvanised	s.j.	soldered joint
g.i.	galvanised iron	sktg	skirting
g.l.	ground level	s.n.	swan neck
g.m.	gauged mortar	s.p.	small pipe
gradg	grading	sprd	spread
HB	half brick	stret.	stretcher
hd/core	hardcore	surf.	surface
hi.	high	swd	softwood
holl.	hollow	swn	sawn
horiz.	horizontal	t. & g.	tongued and grooved
h.r.	half round	tankg	tanking
ht	height	tarmac	tarmacadam
h.w.	hollow wall	th.	thick
hwd	hardwood	tr.	trench
incldg	including	u/c	undercoat
intl	internal	u/grd	underground
jb	jamb	upstd	upstand
jt	joint	vert'l	vertical
L. & C.	level and compact	w.	with
LA	Local Authority	wd	wood
lab.	labour	w.g.f.c.	white glazed fireclay
la.	large	w.i.	wrought iron
l.p.	large pipe	wi.	wide
matl	material	w.p.e.	white porcelain enamelled
max.	maximum	w.s.j.	wiped soldered joint
m.e.	match existing		

Schedules

With the complexity and size of modern buildings (e.g. high-rise developments) there are various aspects of taking-off that involve the measurement of repetitive units of similar characteristics. These are best dealt with by the use of schedules, which are a means of communicating information in an easily readable form on the same document, thus eliminating monotonous cross-referencing and checking between several drawings. These documents are usually prepared by the architect, but may sometimes be prepared by the main contractor to facilitate ordering of materials.

Prime costs and provisional sums

The choice of which term to use should be made in relation to clause A8 of the *Standard Method of Measurement*.

1. *Prime cost sums* – usually intended to cover work/services carried out by nominated sub/contractors or statutory bodies, or goods supplied by nominated suppliers.
2. *Provisional sums* – provided in the Bill of Quantities for work that cannot be accurately measured during the preparation of the tender documents.

These values are later adjusted during the contract period as the actual costs become known. This involves the preparation of a P.C. and Prov. Sum A/C, giving omissions for the previous values followed by additions for the actual costs, plus in each case the respective percentage allowance for profit.

Preambles

These are clauses that are entered at the commencement of each work section of the Bill of Quantities. Their main purposes are to qualify the standard of workmanship and grades of materials to be used in the work. By this means lengthy descriptions may be avoided.

For instance, in concrete work various types of mixes are quoted in preambles stating the respective proportions of cement, sand and aggregate, and the water/cement ratio to be adopted. Then, within the bill descriptions, work may be described as being concrete mix A, B, C etc.

Preambles are usually included during the billing process, after the taking-off and abstracting are completed.

Protection items

Certain of the trade sections of the *Standard Method of Measurement* require the inclusion, as the last bill item, of a clause allowing for the temporary protection of the work completed in that section of the building until such time as the works are completed and handed over, or subsequently covered and hidden by further stages of the work. For example:

Mahogany glazed screen to entrance foyer of building – temporary protection of hardwood framing necessary during successive stages of the work.

SMM protection clauses

Preliminaries Bill

Clause No.
B13(1e)   Protection against inclement weather

Measured Bill

Clause No.
C6   Demolition
D45   Excavation and earthworks
E14   Piling and diaphragm walling
F45   Concrete work
G58   Brickwork and blockwork
H7   Underpinning
J24   Rubble walling
K52   Masonry
L11   Asphalt work
M56   Roofing
N33   Woodwork
P11   Structural steelwork
Q10   Metalwork
R41   Plumbing and mechanical engineering installations
S28   Electrical installations
T35   Floor, wall and ceiling finishes
U19   Glazing
V13   Painting and decorating
W10   Drainage
X15   Fencing

Girthing

One of the initial calculation problems to be overcome in taking off quantities is how to determine the girth (centre-line perimeter measurement) of the external walls or shell of the building. This may be determined in one of two ways:

1. Calculate the perimeter along the face of the wall and then adjust for either over- or under-measure, depending on whether internal or external dimensions are used in the initial build-up; see Example 1.
2. Work around the building on the 'in and out' method, making the corner adjustments as the calculation proceeds; see Example 2.

Example 1

1st method

2/6.000	12.000
2/3.000	6.000
Add for corners: 4/2/0.112½	0.900
Girth =	18.900 m

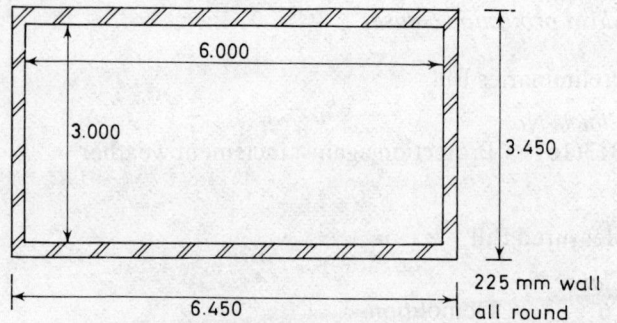

2nd method

2/6.450	12.900
2/3.450	6.900
	19.800
Ddt for corners: 4/2/0.112½	0.900
Girth =	18.900 m

Example 2

In and out method

Wall thickness	0.225
	5.790
Corner allowance	0.225
	5.790
Corner allowance	0.225
	2.440
Corner allowance	0.225
	1.830
5.640 less 0.225	5.415
Corner allowance	0.225
	3.050
Corner allowance	0.225
2.290 less 0.225	2.065
	0.910
Girth =	28.640 m

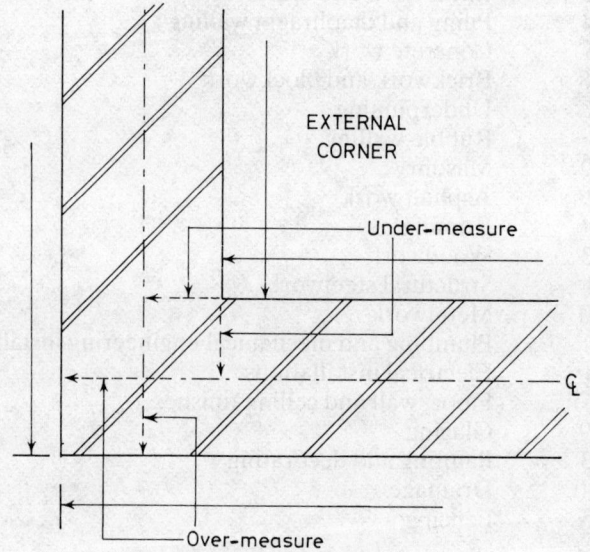

Check:	5.640	
	2.440	
2/	8.080	16.160
2/	5.790	11.580
		27.740

Add for balance of external corners:

Extl	6	4/2/0.112½	0.900
Intl	2		
Balance	4	Girth =	28.640

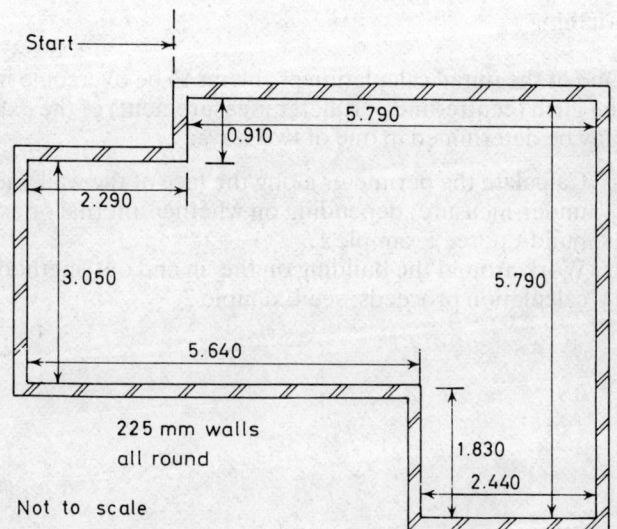

2 Measurement of excavations and foundations

Covered by the rules of measurement laid down in the *Standard Method of Measurement of Building Works*, sections D, F, G and L.

As most building contracts are either a collection of smaller buildings or one large complex structure, the measurement procedure should be related to a series of separate buildings or easily recognised sections of the larger structure. Within each section of the building the surveyor should try to follow the sequence of construction. This will provide an ideal outline for the measurement procedure, with less likelihood of any items being omitted.

The schedule on page 00 gives a general indication of the common items of work measured in the excavation and foundation section of the Bill of Quantities, and at the same time gives the corresponding clauses of the *Standard Method of Measurement*, their implication and the respective units of measurement.

General comments

Before working through the examples there are some aspects of measurement which require a fuller explanation than that already covered by the previous references to the *Standard Method of Measurement*.

1. Excavation of foundation trench

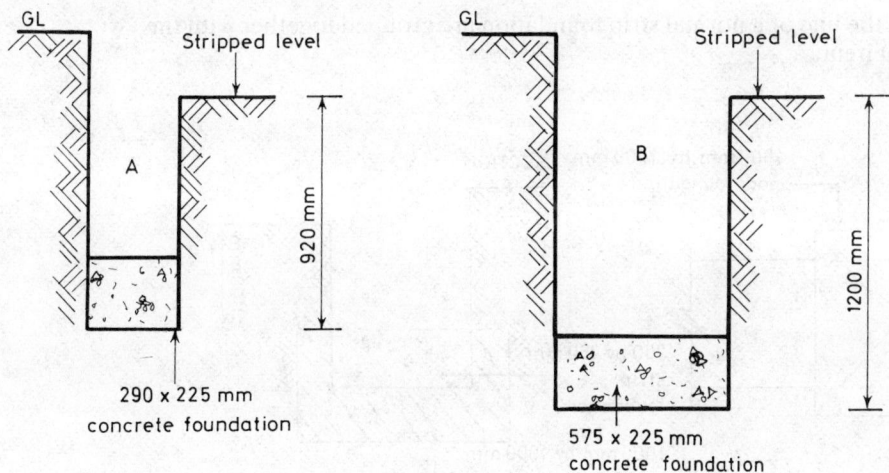

290 x 225 mm
concrete foundation

575 x 225 mm
concrete foundation

Not to scale

Assuming a standard length of trench of 5.00 m, the measurement of trenches A and B would be as shown above.

13

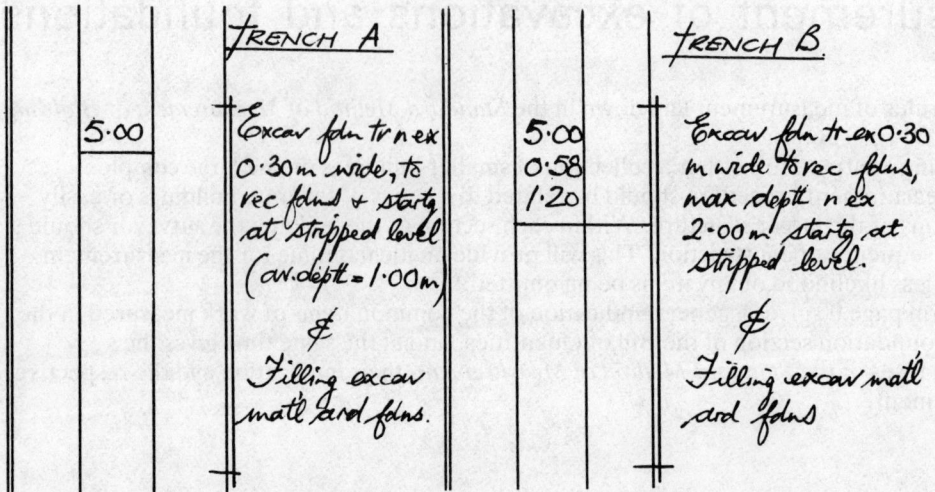

| 5·00 | TRENCH A

Excav fdn tr n ex 0·30 m wide, to rec fdnd & startg at stripped level (av.depth = 1·00m)

&

Filling excav matl and fdns. | 5·00
0·58
1·20 | TRENCH B.

Excav fdn tr ex 0·30 m wide to rec fdns, max depth n ex 2·00 m startg at stripped level.

&

Filling excav matl and fdns |

Ref. SMM D13(6a). Trenches n. ex. 0.30 m wide measured in metres, stating average depth to nearest 250 mm.

D13(6b). Trenches ex. 0.30 m wide measured in cubic metres.

2. *Excavation for pits*

Those that occur on the line of a normal strip foundation are grouped together with the measurement of that item.

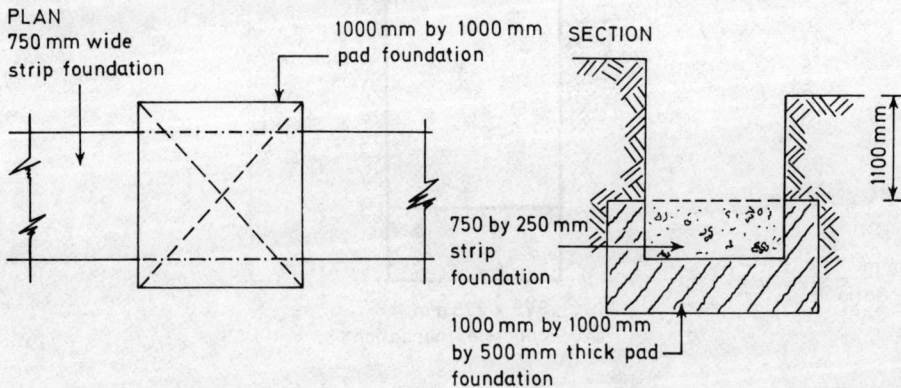

PLAN
750 mm wide strip foundation

1000 mm by 1000 mm pad foundation

SECTION

1100 mm

750 by 250 mm strip foundation

1000 mm by 1000 mm by 500 mm thick pad foundation

14

Ref. SMM D13(6). The measurement procedure for the excavation work would be as shown on the right, assuming a standard length of trench of 5.00 m.

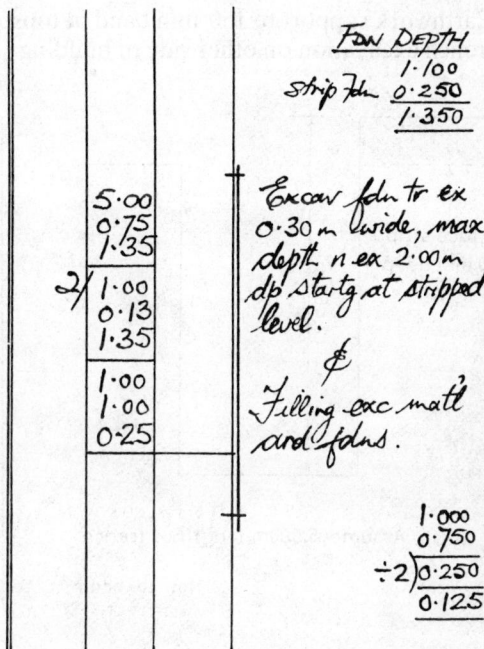

FDN DEPTH
        1·100
strip fdn 0·250
        1·350

	5·00	Excav fdn tr ex
	0·75	0·30 m wide, max
	1·35	depth n.ex 2·00m
2/	1·00	dp startg at stripped
	0·13	level.
	1·35	&
	1·00	Filling exc mat'l
	1·00	and fdns.
	0·25	

        1·000
        0·750
÷2) 0·250
        0·125

Assuming the same overall size for an isolated pit excavation, the measurement procedure would be as shown on the right.

Ref. SMM D13(5). Where both dimensions on plan are less than 1.25 m (including allowances for working space), these pits are measured separately.

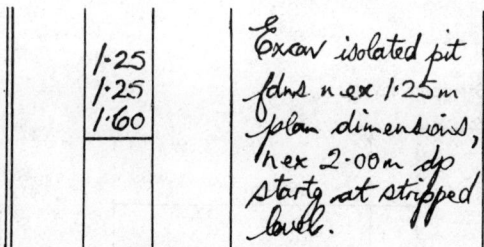

1·25	Excav isolated pit
1·25	fdns n.ex 1·25 m
1·60	plan dimensions,
	n.ex 2·00m dp
	startg at stripped
	level.

3. Earthwork support

To the sides of foundation trenches measured as maximum depth n.ex. 0.25 m; n.ex. 1.00 m; n.ex. 2.00 m; n.ex. 4.00 m deep etc. in further 2.00 m stages as Clause D11. Also classified by the distance between the opposing faces of the formwork, i.e. n.ex. 2.00 m; ex. 2.00 and n.ex. 4.00 m; ex. 4.00 m.

Earthwork support to 150 mm band of topsoil, opposing similar support on opposite external face of trench excavation on other side of building.

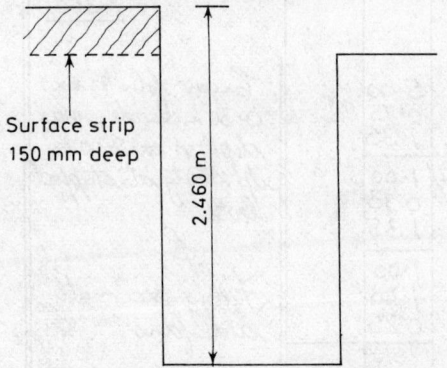

Surface strip
150 mm deep

2.460 m

Assume 5.000m length of trench

Not to scale

$$\begin{array}{r} 2\cdot 460 \\ 0\cdot 150 \\ \hline 2\cdot 310 \end{array}$$

2/ 5·00
2·31

Earthwk support/ n ex 2·00 m between oppos'g faces + max depth n ex 4·00 m dp.

5·00
0·15

Ditto ex 4·00 m between oppos'g faces + max depth n. ex 0·25 m dp. (Topsoil)

Earthwork support to sides of basement (same rules as for support to trench excavations).

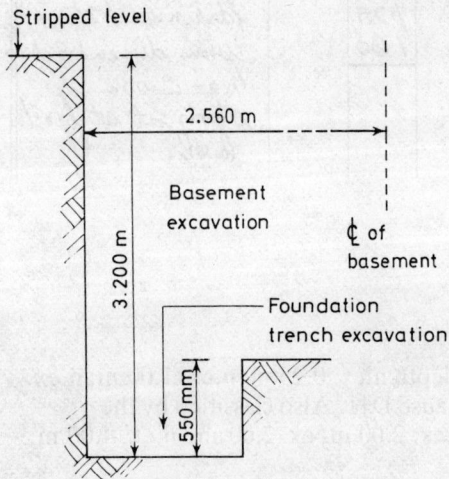

Stripped level

2.560 m

Basement excavation

₵ of basement

3.200 m

Foundation trench excavation

550 mm

Assume 5.000m length of trench

Not to scale

5·00
2·65

Earthwk supports 2·00 to 4·00 m dist between oppos'g faces + max depth n ex 4·00 m dp.

2/ 5·00
0·55

Ditto n ex 2·00 m between oppos'g faces + max depth n. ex 1·00 m dp.

Ref. SMM D11 = trench excavation in basement ex. 250 mm and n.ex. 1.00 m deep.

4. Stepped foundation

On a sloping site an increased thickness of the foundation concrete occurs at the step. As foundation concrete is measured and classified in thickness in accordance with clause F5, an adjustment must be made for this in the actual measurement of the work.

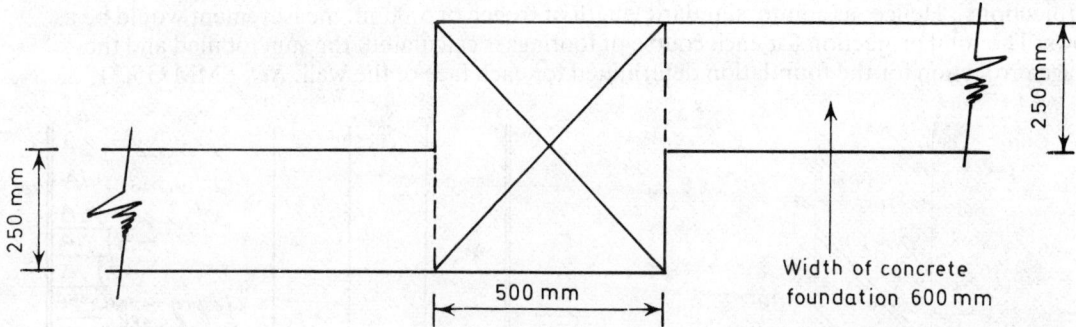

Note. Reversal of normal adjustment, owing to the necessary deduction of previously measured concrete in foundation.

500 mm

250 mm

250 mm

Width of concrete foundation 600 mm

| 0·50 |
| 0·60 |
| 0·25 |

Dđt
Conc (1:2:4/20 mm Agg) in tr folns 150 to 300 mm th. (STEP)
&

Add
Filling excav matl and folns.
&

Dđt
Remove spoil from site.

| 0·50 |
| 0·60 |
| 0·50 |

Conc (1:2:4/20 mm Agg) in tr folns ex 300 mm th. (STEP)
&

Dđt
Filling excav matl and folns.
&

Remove spoil from site.

17

5. Projections

For projections on the main wall thickness (i.e. footings, attached piers etc.) the main wall thickness is measured through the full height of the wall.

Brickwork in footings beyond the main faces of the wall is measured separately and described as in 'projections'. Hence, assuming standard length of trench of 5.00 m, measurement would be as follows. The total projection for each course of footings is calculated, the sum totalled and the average projection for the foundation determined for each face of the wall. *Ref.* SMM G5(3).

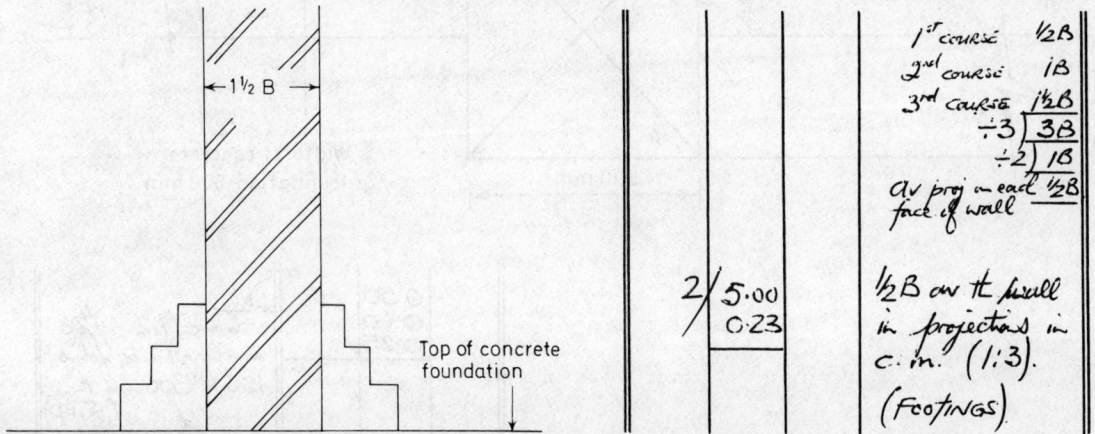

Top of concrete foundation

1½ B

```
1ˢᵗ course    ½B
2ⁿᵈ course    1B
3ʳᵈ course   1½B
        ÷3 ) 3B
        ÷2 ) 1B
av proj on each   ½B
face of wall
```

```
2/ 5.00
   0.23
```

½B av th wall in projections in c.m. (1:3).
(FOOTINGS)

Brickwork in attached piers is again measured as the projection beyond the main wall face, but measured as a lineal item stating the width and depth. Assume 900 mm overall height of brickwork from top of concrete foundation to DPC level. *Ref.* SMM G5(4).

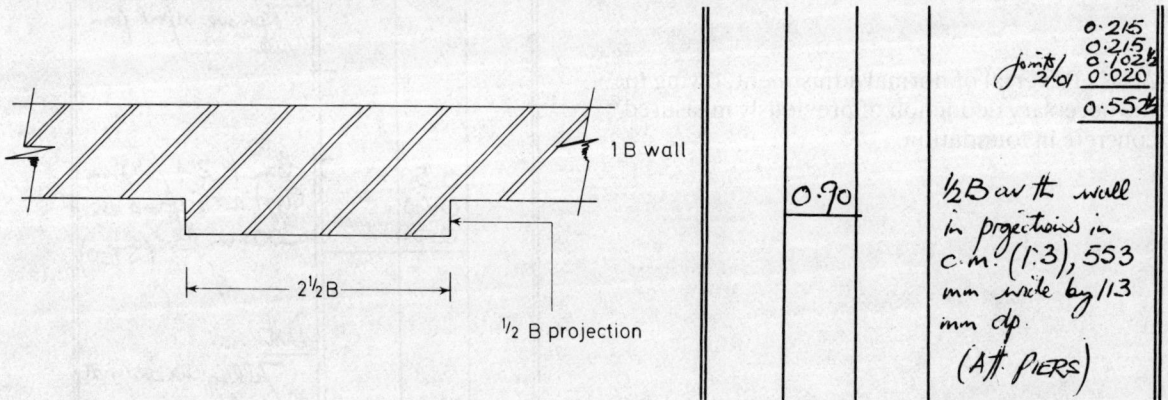

1 B wall

2½B

½ B projection

```
        0.215
joints/ 0.215
2/d     0.102
        0.020
        0.552
```

0.90

½B av th wall in projections in c.m. (1:3), 553 mm wide by 113 mm dp.
(Att. PIERS)

6. Intersections of internal and external walls

The normal approach with excavation, foundations and walls is to start with the measurement of the work related to the external walls first, followed by internal walls, where these vary in thickness, take thicker walls first working through to the thinner ones.

18

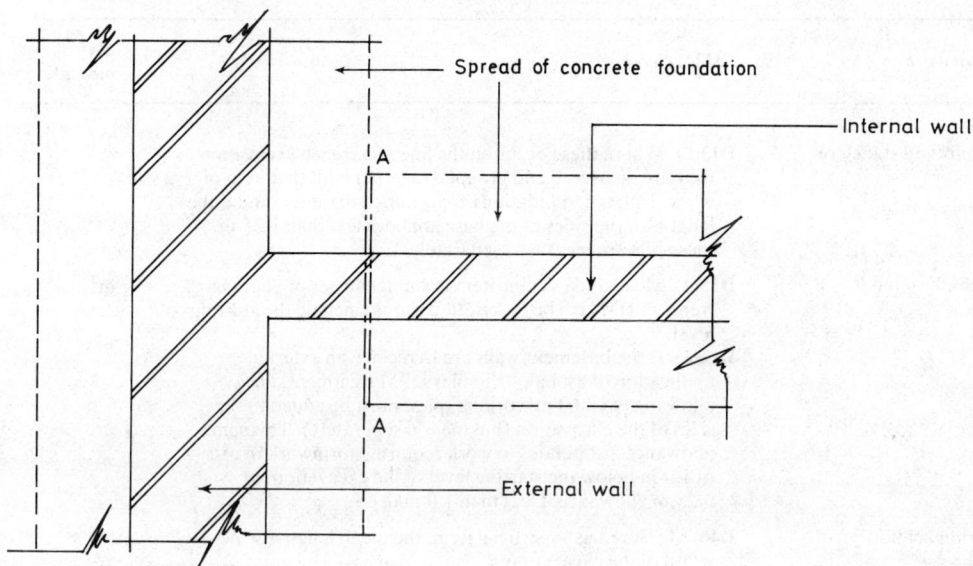

Spread of concrete foundation

Internal wall

External wall

The foundations to the internal walls are then measured between the internal faces of the main trench excavation on opposite sides of the building. A similar procedure is repeated for the walls, the length of the internal walling being taken between the internal faces of the external walls on opposite sides of the building.

This junction between internal and external walls also involves an adjustment for the earthwork supports measured to the main trench excavation, i.e. along line AA on diagram.

Schedule of SMM requirements for substructure work

Item No.	Description of work	SMM reference	Unit of measure
1	Excavating top soil.	**D9**. Measured as a superficial item stating average depth of excavation with the description.	m²
		D28. Separate item measured in relation to the method of disposal of the excavated spoil, e.g. deposited on site in spoil heaps or spread and levelled on site; the distance between the point of excavation and deposition of the material should be stated.	m³
2	Excavating to reduced levels.	**D13(3)**. Measured as a cubic item irrespective of thickness.	m³
3	Excavating foundation trenches and backfilling excavated material into trench. *Note:* it is assumed that all excavated material is to be backfilled around the foundations; then subsequent deductions are made from this total for the volumes of materials incorporated in the foundation construction.	**D13(6)**. Measured in stages of depth, as D11, the description including the level at which excavation commences. Measured as a cubic item taking the width of the proposed foundation as the width of the trench excavation, *except* that where a trench does not exceed 0.30 m wide, this is measured as a lineal item.	m³ m

Item No.	Description of work	SMM reference	Unit of measure
4	Excavating pits and stanchion bases.	**D13(5)**. Where these occur on the line of a trench excavation, they are measured and grouped together with that item of work. Isolated foundations are grouped together, and if the lengths of the sides of the base are both less than 1.25 m, these pits are measured separately.	m³
5	Excavation for basement construction.	**D13(4)**. Measured as cubic item in similar stages of depth as trenches (D11). The description should include the starting level. **D12(1)**. If the basement walls are to receive an external application of asphalt or similar DPM treatment, an allowance of 0.60 m working space must be added to the sizes of the excavation (but meas'd separately). This same allowance also relates to work requiring formwork in excess of 1.0 m below the starting level of the excavation, *or* 0.25 m allowance if less than 1.0 m deep.	m³
6	Level and compact bottom of excavation	**D40**. Measured as superficial item; the mean length by the width of the excavation.	m²
7	Earthwork supports to the sides of trench and basement excavations.	**D14–D24**. Measured in stages of depth as D11, taken as twice the mean length of trench by the depth of the excavation. Unless both sides of the trench start at the same level, each side is measured separately. Not measured unless in excess of 0.25 m deep. *Note:* also classified in relation to the distance between opposing faces.	m²
8	In-situ concrete in foundations and beds.	**F3**. Separation of concrete work according to type of structure and relative position within the building. **F6(1)**. Generally measured as a cubic item, and classified in thicknesses or cross-sectional area, as F5. **F6(2)**. Trench foundations to include extra thickness/width for column bases on the line of the trench and also attached piers.	m³
9	Concrete in walls.	**F5**. Thickness classification. **F6(12)**. Grouping of concrete walls.	m³
10	Fabric reinforcement.	**F12(1)**. Description should include the quality, size and weight of the fabric. **F12(2)**. Measured as a superficial item with no allowance for laps.	m²
11	Formwork to walls.	**F13**. Groupings and measured as a superficial item to the actual concrete faces being supported, distinguishing between the types of surface finish required to the concrete. **F16**. Formwork to walls: description should include the number of separate surfaces in each item.	m²
12	Walls.	**G5**. Description should include the type of brick, bond, type and mix of mortar. Walls generally measured as a superficial item and classified according to G3 and G5. Brickwork to hollow walls measured in separate skins, involving the calculation of differing mean lengths.	m²
13	Formation of cavity to hollow wall including wall ties.	**G9**. Measured the mean length of the cavity by overall depth from the top of the foundation to the DPC level. Description to include width of cavity.	m²
14	Concrete filling to hollow walls.	**F6(18)**. Measured the mean length of the cavity by overall depth from top of foundation to ground level.	m²

Item No.	Description of work	SMM reference	Unit of measure
15	Horizontal DPC.	**G37(1)**. Description should include type of material, quality, number of layers and bedding material etc. **G37(2)**. Lineal item where not exceeding 225 mm width. Over 225 mm width: measured as a superficial item.	 m m²
16	Mastic asphalt tanking.	**L4(1)**. Description should include quality of material, thickness of application and number of coats. **L4(3)**. Measured as a superficial item to the actual surfaces to be covered, for work exceeding 300 mm wide. Work less than 300 mm wide measured as a lineal item.	 m² m
17	Asphalt internal angle fillet. *Note:*at the time of the measurement of items 8 to 13, a corresponding adjustment for the volume of construction work formed in the foundation must be made for backfilling and removal of excavated spoil from the site.	**L4(10)**. Lineal item, angles and ends included.	m
19	Deduction of previously measured half brick skin of hollow wall in common bricks and addition of half brick wall entirely in facing bricks in stretcher bond in cement mortar (1:3).	**G14(9)**. Half brick walls entirely of facings measured separately.	m²
20	Backfilling of excavated material.	**D35**. Adjustment for backfilling trench around external face of building.	m³
21	Keep excavations free from general water.	**D26**. Measured as an item and usually included in Preliminaries Bill.	Item

Sequence of measurement for simple foundations

1. Excavate vegetable soil.
2. Excavate foundation trench and backfill around foundation.
3. Level and compact bottom of excavation.
4. Earthwork supports to sides of excavation.
5. Plain concrete in foundations and adjustment for backfill and disposal of surplus spoil.
6. Brickwork – hollow wall – separate skins.
7. Formation of cavity in hollow wall.
8. Fine concrete filling to cavity.
9. Adjustment for backfill and disposal of surplus spoil for brickwork in foundation.
10. Horizontal damp proof course.
11. Adjustment for facing bricks.
12. Adjustment for backfill around foundations.
13. Hardcore filling below concrete bed.
14. Concrete floor bed.
15. Damp proof membrane.
16 Provision for pumping excavations.

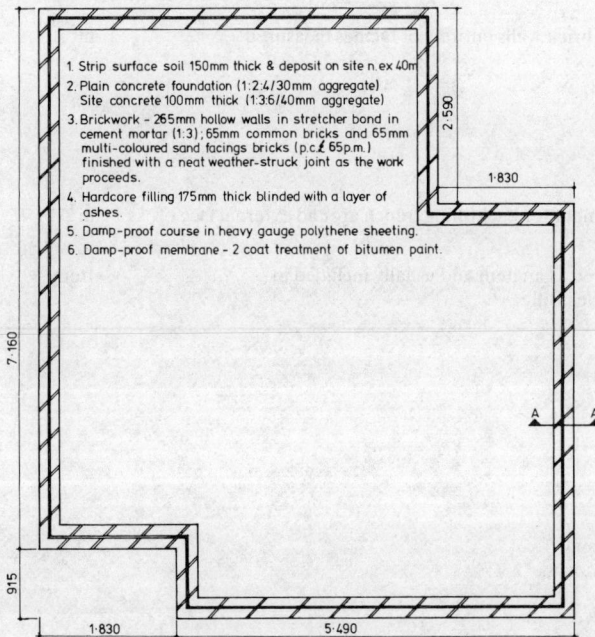

1. Strip surface soil 150mm thick & deposit on site n. ex 40m
2. Plain concrete foundation (1:2:4/30mm aggregate)
   Site concrete 100mm thick (1:3:6/40mm aggregate)
3. Brickwork - 265mm hollow walls in stretcher bond in cement mortar (1:3); 65mm common bricks and 65mm multi-coloured sand facings bricks (p.c.£ 65p.m.) finished with a neat weather-struck joint as the work proceeds.
4. Hardcore filling 175mm thick blinded with a layer of ashes.
5. Damp-proof course in heavy gauge polythene sheeting.
6. Damp-proof membrane - 2 coat treatment of bitumen paint.

Scale 1:100

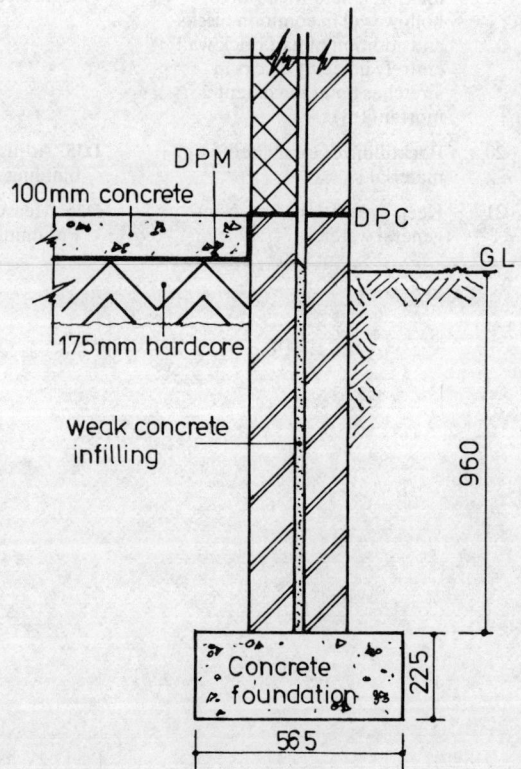

SECTION A-A

Scale 1:20

Panel ② (top)

Build up of foundation trench depth.

Centre line of trench excavation.

CLAUSE D13 (6)–description to include starting level and depth classification as CLAUSE D11.

All measured as backfill with later deductions for construction materials placed in foundation construction.

CLAUSE D40 – Surface treatment.

Handwritten taking-off:

Depth
GL to top of fdn. 0.960
Well fdn. 0.225
1.185
Less site strip 0.150
1.035

Joint
8.075
7.320
×2) 15.395
30.790
less corners 1.060
29.730

Excav tr. exc 0.30m wide to rec. fdns. max. dp. n ex. 2.0m, starting at stripped level.
&
Filling excav. matl. round fdns.

29.73
0.57
1.04

L & C bottom of excav. of tr. rec. conc.

29.73
0.57

Add corners 29.730 / 2.260
total for excv. 1.365 / 31.990
less corners 29.730 / 2.260
Tot. for excav. 27.470

②

Panel ① (bottom)

FOUNDATIONS

ALL WORK UPTO AND INCLUDING DAMP PROOF COURSE

CLAUSE D9 – excavation of topsoil.

Deductions for 'break ins' into overall area measured initially.

CLAUSE D28 – separate measurement for method of disposal of topsoil.

Handwritten taking-off:

Wth of fdn.
Width of fdn. bearing 0.565
÷2) 0.265 / 0.300 / 0.150

Wth of fdn. 0.150
Width
Lngth 5.490
7.600 1.830
0.915 7.320
8.075 0.300
0.300 7.620
8.375
3/150

Excav top soil av 150 mm dp.

Dott. Ditto as last

Deposit topsoil in temp spoil heaps av 40 m from point of excavn.

Dott. Ditto as last.

8.38
7.62

2.59
1.83
1.83
0.92

8.38
7.62
0.15

2.59
1.83
0.15
0.92
0.15

①

CLAUSE G5 (1 & 3) - classification, type of bricks, mortar and bond.

CLAUSE G9(1) - measured mean length by height of brickwork, stating number of wall ties per m².

CLAUSE F6 (18) mean length by height from GL to top of foundation, with thickness as CLAUSE F5 (2).

④

CLAUSE D14 to 18 - measured in depth stages as CLAUSE D11.

CLAUSE D18 - separate measure of earthwork support to external face of trench, as this lies below the face of support to surface strip.

CLAUSE F4 (6) classification as concrete poured against faces of excavation.
CLAUSE F6 (2) classified also according to position, with thickness as CLAUSE F5 (2).

Adjustment of excavated spoil.

③

(5)

29.73 0.28 0.81	<u>Ddt</u> Filling to excavate multi and fills & <u>Add</u> Remove spoil from site.

Adjustment of excavated spoil for brickwork constructed in excavation.

2/ 29.73	Horiz DPC 102¾mm wide of one layer of polythene multi to BS743, lapped 100mm at jts and bedd in c.m. (1:3) (Measd Net)

$$\frac{\text{Facing in Exte}}{\text{Ebs-j}}$$
2 courses above G.L.
and 1 course below.
= 75 × 3
= 0.225

CLAUSE G37 (1 & 2) – not exceeding 225mm therefore lineal item. NB. Use of centre line measurement.

30.38 0.23	<u>Ddt</u> HB skin to holl wall a.b.d. & <u>Add</u> HB skin of holl wall in 65mm multi- col sand fcg brks (PC £65 p.m.) in flat bond, in c.m (1:3) and tly in as work proceeds jt as mrk

Deduct previously measured common brick-work in external skin and add replacement by skin in facing brickwork.

CLAUSE G14 (9) – HB wall entirely of facing bricks.

(6)

```
          8.375
          7.620
      x2) 5.995
          3.990
less excavd  7.150   0.600
                    31.390
```

3.39 0.24 0.15	Filling topsoil from spoil heaps and filled.

```
            Length
            7.160
            0.915
            8.075
less 2/1.265  0.530
            7.545
            Width
            5.490
            1.830
            7.320
            0.530
            6.790
```

7.55 6.79	Hdll/core filling in making up levels to U/S of flrs av 175mm th, spread, lev and compctd and blinded w. a layer of ashes.

CLAUSE D36-superficial item less than 250mm thick, description to include treatment of hardcore surface.

7.55 6.79 0.10	Plain conc (1:3:6/ 40mm agg) to floor n.ex 100mm thick bed on top of hdll/core bed.

CLAUSE F6 (8) – cubic item.

CLAUSE F9 (3)—Treatment to surface of concrete bed measured separately. Measured in compliance with painting section of SMM (Section V) CLAUSES V3 and V4.

Deductions for 'break ins' into main area.

CLAUSE V3 – measured as a lineal item, not exceeding 300 mm girth and given in 150mm stages.

CLAUSE D26— measured as an item, included and priced in the 'preliminaries bill'.

7.55 6.77	Top of conc bed fin to rec DPM & eff screed &	
	Clean surf of conc and apple 2 cts of suitable bituminous waterproofing paint	
2.57 1.83 1.83 0.92	Ddt Ditto as last.	
Add 4/225	29.730 1.060	28.670
28.67	Clean surf of bwk and apple 2 cts of bit paint a.b.d to vert faces of bwk (n.ex 150mm girth).	
Gen	Make good for leaks ex cavl line from general water.	

(7)

Sequence of measurement for basement excavation

1. Excavate vegetable soil, and level and compact bottom of excavation.
2. Excavate basement – maximum depth not exceeding 4.00 m.
3. Excavate foundation trench starting at basement level.
4. Item – provisional sum for excavation and formation of ground beam.
5. Earthwork supports to sides of basement.
6. Earthwork supports to sides of foundation trench.
7. Oversite concrete – in 75 mm bed.
8. HB walls in sec. eng. bricks in lining to face of trench excavation.
9. Horizontal mastic asphalt tanking.
10. Vertical mastic asphalt tanking.
11. Internal angle fillet.
12. Concrete in foundations exceeding 300 mm thick.
13. Fabric reinforcement in foundations.
14. Loading coat of concrete in bed 100–150 mm thick.
15. Fabric reinforcement to last.
16. Concrete in basement walls 150–300 mm thick.
17. Fabric reinforcement to last.
18. Formwork to internal face of wall (smooth face).
19. Formwork to external face of wall (surface left struck).
20. Vertical mastic asphalt tanking.
21. Internal angle fillet.
22. HB walls as outer lining of basement.
23. Adjustment on last for facing bricks.
24. Horizontal damp proof course exceeding 225 mm wide.
25. Adjustment for over-measure of L and C.
26. Adjustment for backfill around outside of basement.
27. Pumping excavations.

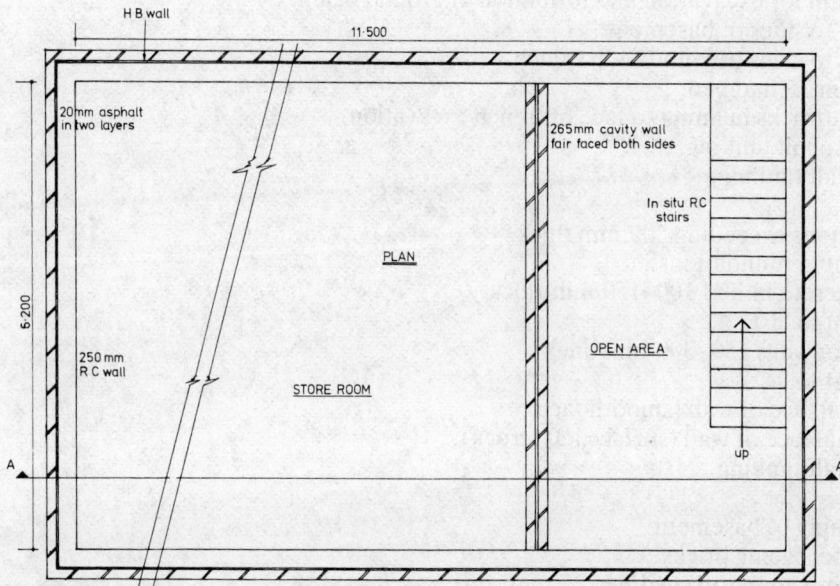

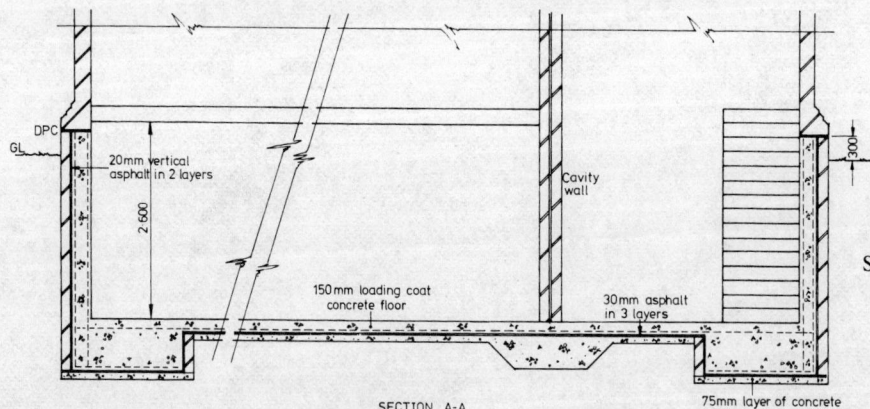

PLAN

STORE ROOM

OPEN AREA

H B wall

11·500

6·200

20mm asphalt
in two layers

265mm cavity wall
fair faced both sides

In situ RC
stairs

250 mm
R C wall

up

A   A

SECTION A-A

DPC
GL

20mm vertical
asphalt in 2 layers

2 600

Cavity
wall

300

Scale 1:100

150mm loading coat
concrete floor

30mm asphalt
in 3 layers

75mm layer of concrete

Walls - 1. H.B. Walls in 65mm second engineering bricks in cement mortar 1:3
           2. Cavity wall in 65mm multi-coloured sand facings in cement mortar
           3. R.Concrete wall (1:2:4 - 40mm aggregate) reinforced with two layers of fabric reinforcement to B S 4483

Floor / Foundation - Reinforced with fabric reinforcement to B S 4483

Foundation - to cavity wall - allow the provisional sum of £400 for all work and materials in formation of ground beam

Page ② (right half):

Excav basmt n ex 4·0m dp strtg 150mm below GL.
&
Remove spoil from site.

CLAUSE D13 (4) - measured to outside of foundations in depth classifications as clause D11.

Work's Space
12·245
6·945
×2)19·190
38·380
4)·600 2·400
40·780

Excav working space to basmt area 4·0m dp max depth strtg at 150 mm below GL and backfill after completion of works.

Excavate working space to basement not exceeding 4.00m deep max depth, starting at 150mm below GL, and backfill after completion of works.

0·102½
0·020
0·800 (SCALED)
0·922½
12·245
6·945
×2)19·190
38·380
tot of both 7·380
31·000

12·25
6·95
2·40

40·78
0·60
2·40

②

Page ① (left half):

BASEMENT

ALL WORK UP TO AND INCLUDING DAMP-PROOF COURSE.

Build-up of overall sizes of excavation, but allowance for working space as CLAUSE D12 (1) assuming external application of asphalt tanking is allowed on each face of the basement construction. N.B. Working space and backfilling measured separately.

Length
Bast 0·102½
Apr 0·020
Conc 0·250
0·372½
11·500
Odd 2/·372½ 0·745
Work's Space)2·245
2)/·600 13·445

Width
6·200
0·745
6·945
1·200
8·145

Excav top soil av 150 mm dp.

CLAUSE D9 - excavation of topsoil.

Dep. top soil in temp spoil heaps av 40m from point of excav.

Less
4/s first u/s bof DPC-100
DPC-#GL 300-0·400
2·600

Add
Conc ·450
Conc ·030
tot of conc 075-0·255
2·455

CLAUSE D28 - separate measurement for method of disposal of topsoil.

13·45
8·15

13·45
8·15
0·15

①

Page ④

CLAUSE D (17) - classification of width between opposing faces and measured total depth inclusive of support required to face of foundation trench excavation. NB. No measure of additional faces formed by excavation for working space.

Earthwork support measured to inside face of trench excavation - ie. in excess of 250mm deep - CLAUSE D15.

Earthwk supports n-ex 4·0m between opposg facg & max depth n-ex 4·0m dp

2/ 12·25	2·46
2/ 6·95	2·46

Outer Face   Inner Face
12·245   38·380
6·945 /ow   ½/
x 2/ 19·190   1/65   14·800
38·380   23·580

Earthwk supports n-ex 2·0m between opposg facg & max depth max 1·0m dp (Fth ti - Basm'ed) INNER FACE

23·58	0·50

Ditto as last but debbs facg of excav 2·455m V-high and n-ex 2·0m between opposg facg v-max depth n-ex 1·0m dp (Fth ti - Basm'ed) OUTER FACE

38·38	0·50

13·445 / 1·200
12·245 / 1·200
8·145 / 1·200
6·945

L & C bottm of basm'nt excav to Re-conc.

12·25	6·95

Page ③

Trench excavation below lowest level of basement excavation, Note, starting level must be stated.

31·00	
1·85	
0·50	

Excav for Itr ax 0·30m width & max depth n-ex 1·0m db starts at 2455 mm below GL (Depth SCALED)

Remove spoil from site

Addit working space
Dept 0·500
less ffconc 0·075
       0·425

40·78	
0·60	
0·43	

Excav working space a.b.d. n-be 4·0m max depth v-ditto.

Allow to Fou. Sum of 2·400 for all work to matter ? formation of grd based under fact to hill wall.

Earthwk Support
Dept 4½ ft to GL 2·600
       0·400
       2·200
Add Conc 150
Cap ·030
Conc ·075 0·255
       2·455

Item

Page 6

Dimension column entries (left):

Mastic Asph to BS 1418
in horiz Dpm to Tanks
in 3 cts to total th
of 30mm on conc bed
& late covered w sac.

	0.051¼
	0.020
	0.071¼
4/2/0.71¼	0.570
Oddment	24.570
	25.160

1/225
6.95

Mastic asp. a.b.d.
in vert Dpm to Tanks
to total th of 20mm
on HB skit nibbly
coved arrises

Ddt	25.160
1/600	0.200
	25.360

25.16
0.50

Fill angle fillet in
2cts th 50mm jnt

	11.500
	6.200
×2)	17.700
	35.400
less angles	4/2/500 4.000
R.C. th. jnt	31.40

25.36

R. conc (1:2:4/40mm
agg) in th bed 300mm
th (bricked agnst cap
shafts). (Dims
scaled)

3.40
1.55
0.65

Clause notes (right):

CLAUSES L3 & L4 - grouping and quality of work included within description.

CLAUSE L5 (1) coved arises included in description as they are sub-sequently covered.

CLAUSE L4 (10) - junction between vertical and horizontal asphalt lineal item inclusive of ends and angles.

CLAUSES F3 & 5 classification of thickness.

Page 5

Dimension column entries (left):

R. conc (1:2:4/20mm
agg) in bed n ex
100mm th laid directly
in contact in the earth

12.25
6.95
0.08

Fin surf of unset
conc to rec app
to tkg.

	11.500
	6.200
×2)	17.700
	35.400
less conrd	
conc th. scaled	4/2/1.30 10.400
Both faces	25.000
Beds scraped	4/1.028 0.410
Gint -	24.570

12.25
6.95

HB skin in sec. bng
Bks in c.m. (1:3½
laid in stret bond.
(Inside face
of trench).

	25.000
Add	0.410
4/1.028	25.410

24.59
0.50

R. o. jt of bwk to
form key for asp.

25.41
0.50

Clause notes (right):

CLAUSE F5(2)- Classification of thickness i.e. not exceeding 100mm thick.

CLAUSE F9 (3)- labour on unset concrete surface measured separately as a superficial item.

CLAUSE G40 - keying of brickwork to receive asphalt measured superficial.

Page 8

		6·200
		11·500
35·40		x2)17·700
2·50		35·400

Junction to wall with face of conc backward wall surface in a thwart face.
Amount (4 No. L. faces).
(Not scaled)

CLAUSE F13 (1)-categories of measurement and -
CLAUSE F13 (2) - measured as actual surface of concrete cast.
CLAUSE F16 - description should include the number of separate surfaces.

		35·400
	Deduct	
	4/2/·250	2·000
		37·400

37·40		
3·15		

Junction to vertical with the of wall surface left struck in to be beads @ 300 mm cts forming stop for asb.
(4 No. surfaces)

Clause F13 - distinguish between wrot and sawn formwork in relation to type of surface finish required to concrete surface.

	Add	37·400
	4/2/·020	0·160
		37·560

Junction between main horizontal and vertical tanking.

37·76		

Ditto asphalt angle fillet a.b.d.

Calculation of centre line of the protective HB outer skin to the asphalt tanking. Kept separate due to increased labour content necessary for 'grouting up' between brickwork and asphalt.

CL of HB Protective skin

		11·500
		0·745
		2·245
		6·200
		0·745
		6·945
		12·245
	x2)19·190	
		38·380
	Less conc	0·410
41·02k		37·970

(8)

Page 7

3·40			
1·55			

BRC fabric reinf. to BS 4483, type A252 laid in floor to 100 mm layers, weight 3·95 kg/m², in 10kg ring wire and stock blocks.
(Measd Net)

	Lgth	Width
	1·500	6·200
2/1·25	2·500	2·500
scaled	9·000	3·700

CLAUSE F 12 - classification and Net measurement.

9·00		
3·70		
0·15		

R. conc. a.b.d. in bed (to slab), 100-150 mm th. laid on rep. tanking.

11·50		
6·20		

BRC fabric a.b.d. + ditto.

	6·200	11·500
	0·250	0·250
	6·450	11·750
		6·450
	x2)8·200	
		36·400

Calculation of the centre line of the concrete wall.

36·40		
2·50		

R. conc (1:2:4/40mm Agg.) in walls 150-300 mm th.

2/36·40		
3·10		

BRC fabric a.b.d. but type B1131 to loc.d. in vert. walls w. 100 mm laps, weight 10·9 kg/m².
(Not scaled)

(7)

Utilise wall centre line, over measure of one layer of fabric compensated by under measure of second layer.

40.76		Backfill top soil from
0.60		temp spoil heaps 40m
0.15		from excav.
		(Allow a heap cart
		by hand of bcky).
	2.600	
	0.02½ 0.112½	
	0.100	
	2.487½	
2/ 6.20		HB skin of hollow wall
2.49		in mult- of sand
		bags in start band
		q.i.b.d. (BASEMENT WALL
		girth area)
6.20		Formation level
2.49		commence & fill
		with a clay 4N°
		buitable backfill/m²
		to BS 1243.
item		Make good to
		facing excavation free
		from ground water.

CLAUSE G5 (3) – skins of hollow walls.
CLAUSE G14 (9) – walls built entirely
of facing bricks.

37.97		HB outer lining of basement
3.15		in sec bag bricks i.c.m.
		(1:3) in short band,
		against vert. dpc.
	from GL 0.300	
	below GL 0.075	
	0.375	
37.76		Rough cutting on sec bag
		bricks for transfer 50km
		grit against with cup
		angle fillet.
37.97	Ddt	HB outer lining to
0.38		deduct a.b.d.
	&	
	Add	
		HB skin in mult-
		of sand bag bricks in
		start band in cm (1:3)
		fin in neat w/struck
		jt as work proceeds
	0.102½	
	0.020	
	0.250	
	0.372½	
	11.500	
	6.200	
	17.700	
	x2) 35.400	
	4/372½ 1.490	
	Add camed 36.890	
36.89		Horiz DPC is dpc applied
0.37		applied in one layer
		127mm th in bwk
		and osc wall width
		(Measured Net)

Adjustment for measurement of facing
bricks at GL: 300 mm GL to DPC plus
allowance for one course below GL.

Calculation of centre line of overall
wall thickness for measurement of
DPC.

CLAUSE G37 (1 & 2) exceeding 225 mm
in width therefore measured as a
superficial item.

3 Measurement of brickwork and blockwork, floors and flat roofs

Brickwork and blockwork

The general approach to the measurement of walls is to follow the sequence: (a) external walls, (b) internal walls, (c) partitions. This will help to simplify the measurement work and reduce the risk of sections of the work being omitted.

The superstructure of a building is usually measured as if there were a complete absence of doors and windows, with the adjustments for the over-measure of walls being made when the doors and windows themselves are measured.

Schedule of SMM requirements for wall construction

Item No.	Description of work	SMM reference	Unit of measure
1	Walls.	**G4(1)**. Generally measured as a superficial item, the mean length by the average height of the wall, *except* facing brickwork or work finished with a fair face, where the work is measured on the exposed face of the wall. **G5**. Description should include type of brick, bond, type and mix of mortar. **G3**. Classification of brickwork according to the type of structure and relative position within the building. **G5(3)**. Brickwork to hollow walls measured in separate skins involving the calculation of differing mean lengths.	m^2
2	Projections.	**G5(4)**. Brickwork in projections (such as attached piers and decorative bands etc.) grouped together and measured separately as a lineal item; width and depth of the projection to be included in the description.	m
3	Formation of cavity to hollow walls, including wall ties.	**G9**. Measured the mean length of the cavity by the overall height from DPC level.	m^2
4	Eaves filling.	Included in the measurement of general brickwork in compliance with G5.	m^2
5	Rough work.	**G10, 11 and 12**. Rough cutting generally included with the item on which it occurs. Rough cutting for chamfers, chases etc. measured as a lineal item.	m
6	Facing brickwork.	**G14(2 & 3)**. Facing brickwork shall give a full description of the materials, etc., and is measured superficial as extra over the wall on which it occurs.	m^2
7	Fair cutting.	**G14(13)**. Fair cutting generally included with brickwork on which it occurs, measured separately as a lineal item only for curved cutting.	m
8	Facing brickwork.	**G14(9)**. Where walls of a half-brick or one-brick thickness are constructed wholly of facing bricks or are finished with a fair face on both sides, they must be measured separately.	m^2
		G14(6 & 10). Facing brickwork to the reveals of door or window openings and the ends of walls, measured as a lineal item; reveal width included in the description in half-brick stages.	m

Item No.	Description of work	SMM reference	Unit of measure
9	Blockwork in partitions.	**G26**. Description should include type and size of blocks, bond, type and mix of mortar. Classified in accordance with G27 and measured as a superficial item.	m²
10	Damp-proof courses.	**G37(1)**. Description should include type of material, thickness and quality and type of bedding material.	
		G37(2). Damp-proof courses not exceeding 225 mm width measured as lineal item; in excess of 225 mm width measured as a superficial item.	m m²
11	Airbricks.	**G52**. Formation of holes for the incorporation of air bricks, and the air bricks themselves, measured separately as enumerated items; sizes included in the description.	No.
12	Protection.	**G58**. Given as an item.	Item

Floors and roofs

Covered by the rules of measurement laid down in the *Standard Method of Measurement of Building Works*, sections D, F, G, L, M, N, P and V. This work breaks down into two sections:

- Floors/flat roof construction (see below).
- Pitched roof construction (see Chapter 4).

The work in floors and flat roofs has two aspects: (a) structural floor/roof; (b) floor/roof finish – although in the case of floors the finishes, where possible, particularly with solid floors, are frequently scheduled and measured in conjunction with all internal finishes.

In the case of floor construction it is common practice to work through the building floor by floor, either starting at the bottom of the building and working upwards or alternatively in the reverse direction.

The work can be loosely classified under the heading of timber and concrete construction, or as ground floors and suspended floor/roof construction.

Timber construction

Determination of number of joists. Clear length between wall surfaces *less* 2 × [end allowance for air space (between end joist and wall surface), and half the joist thickness]. Resultant length (AB) is divided by recommended spacing for the joists, then rounded up to the next whole number plus one to give number of joists required.

With upper floors, the construction will have to incorporate joists of differing thicknesses where it is necessary to trim around openings such as chimney breasts, stair-wells etc.

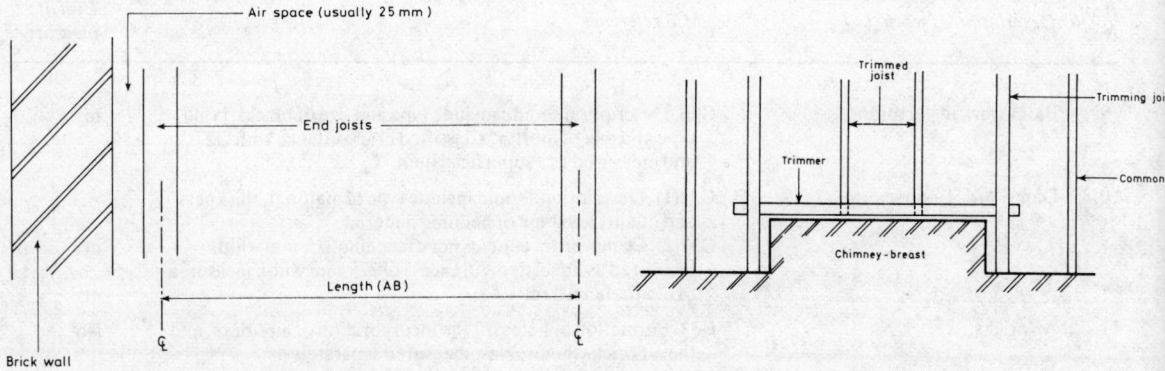

Adjustment for trimming around opening. Complete floor is measured initially as all common joists, followed by:

1. A deduction for two full lengths of common joist, and the addition of two lengths of trimming joist.
2. A further deduction for the overmeasured ends of the trimmed joists.
3. Addition of increased-size member as trimmer.

In the example below, the room span for joists including bearings is assumed to be 2.800 metres.

36

Concrete construction

The general sequence of measurement with all items of concrete work within a building covers three aspects: (a) formwork, (b) reinforcement, (c) concrete. The main point of emphasis is that formwork is measured to the actual surfaces requiring support.

With floor/roof slabs cast with integral beams (i.e. T or L beams), formwork is measured to the complete floor/roof area as if there were no beams, followed by deduction of areas of formwork to floor/roof soffits where beams exist and then the addition of formwork to sides and soffits of beams.

Note In the example below, B is an attached beam, not exceeding the 3:1 depth/width ratio as per Clause F6(14); therefore the concrete projection below the underside of the suspended concrete slab is measured separately, but classified as concrete in suspended slab. Formwork to the soffit of the suspended floor slab would be measured as to three separate surfaces. *Ref.* SMM F15(1).

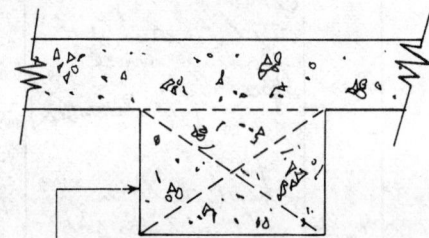

Concrete in beams is measured below the lower soffit of the floor/roof slab

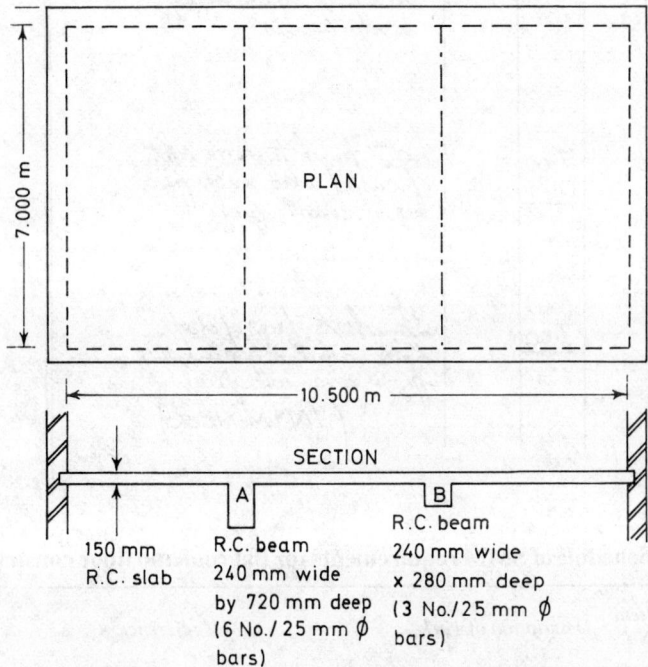

PLAN

7.000 m

10.500 m

SECTION

A

B

150 mm
R.C. slab

R.C. beam
240 mm wide
by 720 mm deep
(6 No./25 mm ∅
bars)

R.C. beam
240 mm wide
x 280 mm deep
(3 No./25 mm ∅
bars)

37

BEAM A

		Bear'gs 2/100	7.000
			0.200
			7.200

	7.20	R.conc (1:2:4/20mm Agg) in deep beam proj'g below U/s of R.C. slab. (c.s.a 0.10-0.25) m²
	0.24	
	0.72	

		less conc	7.000
		over 2/of	0.080
			7.120
		Std Hooks	
		2/12/25	0.600
			7.720

| 2/3/ | 7.72 | M.s. reinf'nt (cranked as per bending schedule) 25 mm ø to BS 4449 in dp beams. |

| | 7.00 | Dott Fmwk to horiz soffit of susp'd slab, surf fin w a smooth face. |
| | 0.24 | |

| | 7.00 | Fmwk to faces of deep horiz attached beam, surf fin w a smooth face. (1 N° MEMBER) |

BEAM B

	7.20	R.conc (1:2:4/20mm Agg) in susp'd slab 100-150mm th.
	0.24	
	0.28	

| 3/ | 7.72 | M.s. reinf'nt (cranked as per bend'g schedule) 25 mm ø to BS4449 in susp'd slab. |

| | 7.00 | Dott Fmwk to horiz soffit of susp'd slab, surf fin w. a smooth face. |
| | 0.24 | |

| | | Fmwk to faces of horiz attached beam, surf fin w a smooth face. (1 N° MEMBER) |

| | 10.50 | Fmwk to horiz soffit of susp'd slab, surf fin w a smooth face. (3 N° SEP SURFACES) |
| | 7.00 | |

Schedule of SMM requirements for flat roof and floor construction

Item No.	Description of work	SMM reference	Unit of measure
	Ground floors and flat roofs in timber construction		
1	Floor/roof joists, trimmers etc.	**N1**. Description to include type and quality of timber, for carpentry work generally described as 'sawn'. Information on preservative treatment, e.g. to ends of joists before fixing, measured as enumerated item.	No.
2	Structural timber members.	**N2**. Classification and grouping of members. Carcassing items generally measured as a lineal item. **N1(6)**. Deals with the measurement of timbers in long continuous lengths, e.g. softwoods exceeding 4.20 m long and hardwoods exceeding 3.00 m long are measured separately in further stages of 0.30 m.	m

38

Item No.	Description of work	SMM reference	Unit of measure
3	Extra labour on structural timber members.	N3(4). Notching, fitted ends and trimming joists around openings measured as enumerated items.	No.
4	Strutting.	N2(2). Measured as a lineal item over the joists for the full floor/roof span. Description should include the sizes of the strutting and the depth of the joists in the construction.	m
5	Boarding and flooring generally.	N4. Measured net with no allowance for extra timber in joints. Description to include details of board thickness, jointing technique and any surface treatment required, including cleaning off on completion.	m²
6	Roof boarding.	N4(1). Classification (flat or sloping). Firrings measured separately as a lineal item stating width and average depth.	m
7	Boarding in gutter construction.	N4(1). Measured separately as a superficial item, including boarding to sides of gutter.	m²
8	Labours on boarding and flooring.	N5(1 & 2). Raking or circular cutting, rebates, grooves etc. measured as a lineal item.	m
		N5(4). Notches and the like are enumerated.	No.
9	Nosings and margins.	N7. Measured separately as a lineal item, stating sizes of section used.	m
10	Protection.	N33. Protection given as an item.	Item

Ground floors and flat roofs in concrete construction

Item No.	Description of work	SMM reference	Unit of measure
11	Concrete work generally.	F4. Description to include type of materials and mix proportions.	
		F5. Size classification. (i) Where concrete placed in members of small cross-sectional area such as beams and columns, work should be described as in one of the following groups: not exceeding 0.03 m²; 0.03–0.10 m²; 0.10–0.25 m²; exceeding 0.25 m. (ii) Where placed in beds or slabs, work should be described as in one of the following groups: not exceeding 100 mm; 100–150 mm; 150–300 mm; exceeding 300 mm.	
		F3. Separation of concrete work according to type of structure and relative position within the building.	
		F6(1). Categories of concrete work – generally measured as a cubic item.	m³
		F6(9). Suspended slabs are measured over the bearings.	
12	Reinforcement generally.	F11 & 12. Description should include the quality, size and weight of fabric. For bar reinforcement description should include the quality, size, bends, hooks and tying wire etc.	
13	Bar reinforcement.	F11. Classification by situation of usage, and separation into straight bars, bent bars, curved bars and links etc. *Note:* this is an exceptional case, being measured as a lineal item but billed by weight (tonnes). (Exceptionally long bars, over 12.00 m long, measured separately in further 3.00 m stages.)	m
14	Fabric reinforcement.	F12. Measured as a superficial item with no allowance for laps.	m²
15	Formwork to concrete.	F13. Measured as a superficial item to the actual concrete faces being supported, distinguishing between the type of surface finish required to the concrete. *Note:* description is deemed to include for easing, striking and removal of formwork.	m²
		F13, 14 & 15. Classification according to situation of usage, positional limitations; e.g. formwork required at excessive heights or isolated positions.	m²

Item No.	Description of work	SMM reference	Unit of measure
16	Formwork to edges of slabs and upstands generally.	**F14**. If exceeding 1.00 m high, measured as a superficial item. If not exceeding 1.00 m high, measured as a lineal item in the following groupings: not exceeding 250 mm high; 250–500 mm high; 500–1000 mm high.	m² / m
17	Protection.	**F45**. Given as an item.	Item
18	Hardcore placed below concrete ground floor slabs.	**D34**. Hardcore filling measured as equivalent of the space to be filled. **D36**. Measured as a cubic item where placed in beds over 250 mm thick, and as a superficial item when in beds not exceeding 250 mm thick.	m³ / m²
19	Dwarf support walls to hollow ground floor construction.	**G5(1 & 3)**. Description should include type of brick, bond, type and mix of mortar. Measured as a superficial item.	m²
20	Protection.	**G58**. Protection given as an item.	Item
	Asphalt roofing		
21	Asphalt work generally.	**L4(1)**. Description to include type of material, thickness of application, number of coats etc. **L4(5)**. Where an underlay is required, this should be included in the description of the asphalt and not measured separately.	
22	Asphalt coverings.	**L4(3)**. Classification and grouping of work. Measured as a superficial item where areas exceed 300 mm wide or as a lineal item for areas not exceeding 300 mm wide (width stated in 150 mm stages).	m² / m
23	Asphalt angle fillets.	**L4(10)**. Measured as a lineal item, angles and ends included.	m
24	Labour on asphalt work.	**L5(1)**. Edges and turning into grooves measured as a lineal item.	m
25	Asphalt upstands.	**L6(1)**. Skirtings, aprons etc. measured separately as a lineal item; description to include face width.	m
26	Asphalt work to gutters.	**L7(1)**. Measured separately from main roof coverings as a lineal item; girth on face to be included in description.	m
27	Asphalt to cesspools and around projections through the roof.	**L8**. Cesspool linings and collars around pipes and the like penetrating a roof surface measured as enumerated items.	No.
	Bitumen felt roofing		
28	Bitumen felt work generally.	**M33(1)**. Description to include type of material, number of layers, laps, method of fixing and any surface treatment. **M33(2)**. Where underlay is required, this should be included in the description of the bitumen felt and not measured separately.	
29	Bitumen felt roof coverings.	**M34(1)**. Classification and grouping of work. Measured as a superficial item where areas exceed 300 mm wide or as a lineal item for areas not exceeding 300 mm wide (width stated in 150 mm increments).	m² / m
30	Labours on bitumen felt roofing work.	**M35(1)**. Cutting, edges and turning into grooves etc. measured as a lineal item.	m
31	Edge treatments.	**M36**. Flashings, skirtings etc. measured as a lineal item, including width or girth in description, in 150 mm stages.	m
32	Gutters.	**M37**. Felt work in gutters measured separately as a lineal item, including width or girth in description.	m

Item No.	Description of work	SMM reference	Unit of measure
	Sheet metal roofing		
33	Metal roofing generally.	**M40(1)**. Description to include type and quality of materials used.	
		M40(2). Standard allowances in relation to the various jointing techniques used in jointing metal sheets.	
		M41. Classification and grouping of work. Measured as a superficial item.	m^2
34	Labours on metal roofing.	**M42**. Raking and circular cutting etc. measured as a lineal item. Bossed ends and the like measured as an enumerated item.	m No.
35	Fixing of metal sheets.	**M43**. Fixings such as copper nailing measured as a lineal item; description to include spacing.	m
36	Cesspools.	**M45**. Cesspool linings measured as an enumerated item.	No.
37	Protection.	**M56**. Protection given as an item.	Item

Sequence of measurement for brickwork to clubhouse above DPC level

The work is measured at this stage assuming a complete absence of door and window openings. The necessary deductions for the over-measure of the brickwork will then be made when the windows/doors themselves are measured.

External walls

1. Walls to ground floor: external H.B. skin in facing bricks, internal skin in 100 mm blockwork, and formation of 62.5 mm cavity.
2. Eaves filling to three walls of ground floor section (added to previously measured general brickwork).
3. Cavity wall at ground level in blockwork separating lounge and toilet areas.
4. Brickwork to first floor: as breakdown for ground floor.
5. Eaves filling to all walls at first floor/roof level.

Internal walls

6. Internal blockwork partitions to toilet area at ground floor level.
7. Ditto to changing rooms at first floor level.

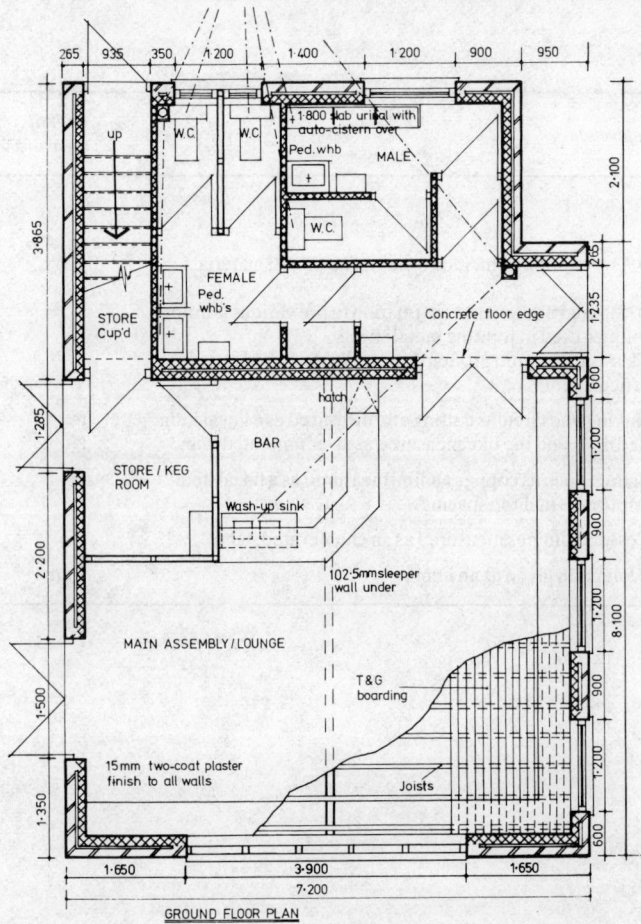

265 | 935 | 350 | 1·200 | 1·400 | 1·200 | 900 | 950

up

W.C.

W.C.

1·800 slab urinal with auto-cistern over

Ped.whb

MALE

2·100

1·235

W.C.

3·865

STORE Cup'd

FEMALE Ped. whb's

Concrete floor edge

600

1·285

hatch

1·200

BAR

900

STORE / KEG ROOM

Wash-up sink

2·200

1·200

8·100

102·5mm sleeper wall under

MAIN ASSEMBLY/LOUNGE

T&G boarding

900

1·500

1·200

15mm two-coat plaster finish to all walls

Joists

600

1·350

1·650 | 3·900 | 1·650

7·200

GROUND FLOOR PLAN

Scale 1:100

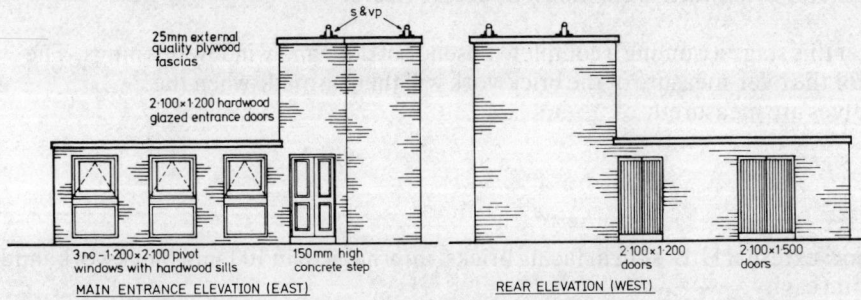

25mm external quality plywood fascias

s & vp

2·100×1·200 hardwood glazed entrance doors

3 no - 1·200 × 2·100 pivot windows with hardwood sills

150mm high concrete step

MAIN ENTRANCE ELEVATION (EAST)

2·100 × 1·200 doors

2·100 × 1·500 doors

REAR ELEVATION (WEST)

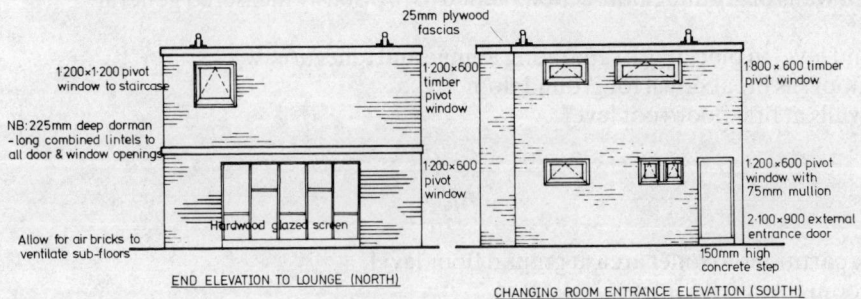

25mm plywood fascias

1·200 × 1·200 pivot window to staircase

1·200 × 600 timber pivot window

1·800 × 600 timber pivot window

NB: 225mm deep dorman - long combined lintels to all door & window openings

1·200 × 600 pivot window

1·200 × 600 pivot window with 75mm mullion

Allow for air bricks to ventilate sub-floors

Hardwood glazed screen

2·100 × 900 external entrance door

150mm high concrete step

END ELEVATION TO LOUNGE (NORTH)

CHANGING ROOM ENTRANCE ELEVATION (SOUTH)

Scale 1:200

42

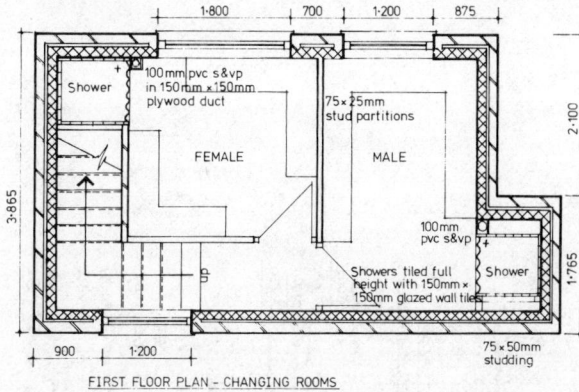

Brickwork 265 mm cavity
walling with approved facings

1·800 700 1·200 875

2·100

Shower

100 mm pvc s&vp
in 150 mm × 150 mm
plywood duct

75 × 25mm
stud partitions

FEMALE

MALE

3·865

100 mm
pvc s&vp

1·765

Showers tiled full
height with 150mm ×
150mm glazed wall tiles

Shower

up

Scale 1:100

900 1·200

75 × 50mm
studding

FIRST FLOOR PLAN - CHANGING ROOMS

Specification notes

Ceilings to be 2.400 m high throughout.
DPCs: Visqueen 1000 under slabs. Ruberoid hyload pitch polymer in walls and openings.
Structural timber to be pressure tanalised.   Plaster: two coats 16 mm plaster to new internal bwk.
Allow for 150 × 150 mm glazed wall tiles to toilet areas and showers.
Provide & fix thermoplastic floor tiles (3 mm) to first-floor & ground-floor toilets and entrance hall.
High roof: 12 mm mineral chippings on 3 layers bitumen-bonded roofing felt on 19 mm weyroc (roof grade) on firrings to falls, 1 in 60 min., on 63 × 200 joists at 450 ctrs with 75 mm thick insulation quilt to joists with 13 mm aluminium foil-backed plasterboard & 3 mm skim finish.
Low roof: 12 mm mineral chippings on 3 layers of bitumen-bonded roofing felt on 22 mm weyroc (roof grade) on firrings 1:60 min., on 'Corply' 12 beams at 600 mm c/cs, with 75 mm thick insulation quilt to joists with 13 mm aluminium foil backed pbd and 3 mm thick skim finish.
Floors: ground: 21 mm polished t&g strip bdg on 175 × 50 joists at 400 ctrs on 100 × 50 wall plates, on DPCs on 102.5 mm sleeper wall on 100 mm oversite concrete (1:3:6) on 150 mm hardcore, 125 mm from top site concrete to underside joists.
         First: 19 mm t&g boards on 175 × 50 joists at 400 ctrs with 25 mm insulation quilt to joists.
Electrician: all work to be done in accordance with current IEE & EMEB Regulations; all pendants operated from master switch panel in store/keg room.
Drainage: s.v.p.s = 100 mm p.v.c.
           w.h.b.s = 32 mm p.v.c.
           showers = 38 mm p.v.c.
           r.w.p.s = 62 mm p.v.c.
Stairs: 900 wide, pitch = 38°,   headroom = not less than 2 m vertically to pitch line.
           14 no. risers × 186 mm   going = 250 mm
           strings = ex. 250 × 38   risers = ex.25 mm thick
Note: Ground floor lobby/w.c.s/store floor to be concrete, i.e. vinyl floor tiles on 50 mm screed on Visqueen 1000 DPM on 100 mm site concrete (1:3:6) on min. 150 mm well-consolidated hardcore.
       Foundations: 600 mm × 225 mm founds (× 450 mm min. cover) with mesh reinforcement to weigh not less than 2.565 kg/m²

①

CLUBHOUSE

BRICKWORK ABOVE DAMP-PROOF COURSE LEVEL ONLY

GROUND FLOOR
EXTERNAL WALLS

```
        8.700
        2.700
        7.200
   x2 │17.400
        34.800
Less cav'd   4/0.2%   0.410
Cl. ext. skin        34.390
                     34.800
Less cav'd  4/2/133%  33.750
Cl. of cavity         34.800
                      1.720
Cav'd    4/2/.25       33.080
Cl. 2nd skin          2.500
                      0.150
· less app/sol   ─────2.350
```

| | | 34.80 |
| | | 2.35 |

HB skin of hell wall
in 65 mm dand fcg
bkes (PC £65 p.m²)
laid in stret bond in
g.m. (1:1:6) and fi-
in bucket handle fcg as
work proceeds.

CLAUSE G5 (1 & 3) – classification, type
of bricks, mortar and bond.
CLAUSE G14 (9) – half brick wall
entirely of facings measured separately.

Formation of cav 62½
mm wide & hell wall in
clay 4No butterfly wall
ties/m² to BS 1243.

| | | 33.73 |
| | | 2.35 |

CLAUSE G9 (1) – formation of cavity
to include type and number of wall
ties per m², measured mean length
by height of wall.

100 mm t. skin of thermalite
blks laid in g.m. (1:1:6)
in hell wall constrn.

| | | 33.08 |
| | | 2.35 |

CLAUSE G26 – description to include type
of blocks, mortar etc.
CLAUSE G27 – measured superficial and
classified as in skins of hollow walls.

②

```
less 2/.265    7.200
               0.530
               6.670
               34.800
less cav'd     0.860
         4/25  33.940
```

1B wall in sand lime
(1:1:6) in English Bond   (EAVES
(to 75 mm).   FILLING)

| | | 33.94 |
| | | 0.18 |

INTL WALL BETWEEN
LOUNGE & TOILET AREA

100 mm t. skin of
thermalite a.b.d.

| | | 2/6.67 |
| | | 2.35 |

Formation of 62½ mm
cav a.b.d.

| | | 6.67 |
| | | 2.35 |

FIRST FLOOR
EXTERNAL WALLS

```
               7.200
               3.865
         x2 │11.065
               22.130
               0.410
               21.720
less cav'd     4/0.2%
Cl. ext. skin  21.720
```

HB skin of hell
wall a.b.d.
                    (Starting pt
                     ht. level
                    (Height Scales)
                     22.130
less cav'd   4/2/133% 1.070
Cl. cav.   21.060

| | | 21.72 |
| | | 2.70 |

CLAUSE G30 – Measured as superficial
item over all joists and added to
general brickwork.

Internal wall at G.F. level between
lounge and toilet area.

④

2/2/ 2·58 ⎯⎯⎯ 2·58	Bonding end of blockwork partn to prev'construct'n with blockwork skin of cav. wall in clay formation where pockets'd l.d and alternate course of blockwork.
1·60 ⎯⎯⎯ 2·58	100 mm tk skin of thermalite a.b.d. (STAIRCASE WALL).
Item	Allow for protection of brickwork and blockwork.

③

N.B. Main lengths of thermalite
partitioning scaled off drawing
and
2.400 m storey height as per
specification notes + average 175mm
additional depth to form adequate
fire stop in fire stopping cavity of
floor and roof construction
= 2.575 total height.

21·06 ⎯⎯⎯ 2·70	formation of 622 mm cav. a.b.d.
20·41 ⎯⎯⎯ 2·70	Ddt 4/215 22·130 1·720 Add it'skin 20·410 100 mm tk skin of thermalite a.b.d.
21·27 ⎯⎯⎯ 0·25	Ddt 4/215 22·130 0·860 21·270 1B. wall in cmns in g.m. (1:1:6) a.b.d. (EAVES FILLING)
	Int'l Block walls at grd floor level Ddt 2/265 3·865 0·530 3·335 NB other lengths scaled.
2/3·34 ⎯⎯⎯ 2·58 2·95 ⎯⎯⎯ 2·58 1·24 ⎯⎯⎯ 2·58 2·00 ⎯⎯⎯ 2·58 1·20 ⎯⎯⎯ 2·58	100 mm tk skin of thermalite blks a.b.d. in int'l walls and partns. (G.F.L.)

Sequence of measurement for ground floor to clubhouse

Suspended timber floor to main assembly lounge

This work is measured complete including all work above oversite concrete, i.e. honeycombed sleeper walls, damp proof courses, wall plates, floor joists and boarding, and air bricks.

1. 102.5 mm thick dwarf walls.
2. DPCs to dwarf walls.
3. Wall plates.
4. Air bricks and formation of holes in brickwork.
5. floor joists.
6. Tongued and grooved strip flooring.

Solid ground floor to toilet area

7. Bed of hardcore including sand blinding.
8. Bed of concrete as ground floor slab.
9. 1000 gauge DPM to underside of floor screed.
10. Sand/cement floor screed.

Sequence of measurement for timber flat roof to clubhouse

1. Wallplates.
2. Softwood roof joists.
3. Roof boarding – firring measured separately.
4. 100 mm thick insulating quilt.
5. Fascia boarding and softwood batten.
6. Softwood triangular fillet to verge treatment at edges of roof.
7. Three layer bituminous roofing felt.
8. Turn down at verge treatment.
9. Turn down at eaves.
10. Upstand of felt of abutment of low roof with brick walls to first floor construction.

CLUBHOUSE (continued)

3/2	225 x 150 mm φ bvf clay airbrk.	CLAUSE G52 – enumerated item with formation of hole in wall measured separately.
	Forming hole i. 265mm bk wall for airbrk size 225x150mm.	
16/6·77	Low carcass tro.&.joists 6·070	Joist calculation gives number of spaces between joists therefore add on one extra to give a number of joists.
	÷400/ 5·970 .100 / 14·9	
	= 15 spaces + 1 = 16 N° joists	
	Less ½ jsts 7·200 0·530 6·670	
	Add bearing onto wallplate 0·100 6·770	
	Duct & joists 50x75mm (2 lengths 6·690 lg 6·960m long) (carcassing Fir butt)	CLAUSE N1 (6) – structural timbers over 4.20m in length measured separately in further stages of 0.30m or alternatively in floor allow lapped joint over centre wallplate and place in two lengths.
6·67 2·07	21mm φ hwd T&G strip flrg i. 100mm width chamfered & secretly fxd to duct jsts w. 1¼0 flr brad at ea. posit. i. clay flooring ready for polishg.	CLAUSE N4 – strip flooring not exceeding 100 mm nominal width – flooring in openings measured separately.

CLUBHOUSE (underlined)

GROUND FLOOR CONSTRUCTION
TIMBER FLOOR to
ASSEMBLY LOUNGE

3/6·07 0·08	½/1 of 3nc & ½ of ½ jsts 0·125 less wallplate 0·050 0·075	
	2/0·600 1·200 3/1·200 3·600 2/0·900 1·800 6·600 0·530 6·070	
	less 2/265	
	HB dbr. long armed deeper reinfd cb.dust support to flr jsts i. 65mm bnds i. c.m. (1:3).	Measured in compliance with normal rules for brickwork.
3/6·07	Horiz DPC 102½mm wide of one layer of Ruberoid Hyload pitch polymer sheetg to BS 743, lapped 100mm at jts. & bedd in c.m. (1:3). (N ould'd N ext)	CLAUSES G37 (1 & 2) – type, quality etc to be stated in description. Not exceeding 225 mm wide therefore measured as a lineal item.
3/6·07	100x50 mm sawn swd wallplate (carcassing fir butt). & Bedg 100 mm wide wallplate i. c.m. (1:3)	CLAUSE N6 (3)–carcassing timbers automatically assumed to be of 'sawn' quality material.

①

CLAUSE D.36 - Hardcore not exceeding 250 mm thick measured as a superficial item.

CLAUSE F6 (8) Beds of concrete measured as a cubic item and classified in thickness as CLAUSE F5 (2).

CLAUSE F9 (3) - labour on unset concrete.

Measured as superficial item as DPC's in compliance with CLAUSE G37.

④

CLAUSE N6 (3) - measured lineal except where less than 300 mm in length then enumerated.

CLAUSE T35 - allowance for protection of completed work, description to include area concerned.

③

5.72		C/s screed (1:3) 50mm thick laid on bed of conc. and trowelled smooth to rec. thermoplastic tiles.
3.34		
1:24		
1:22		

CLAUSE T4 - description of material and composition, and surface finish as CLAUSE T13 (2) and measured separately.

5.72		225 x 225mm Vinyl flr tiles laid on felt spread on proprietary adhesive.
3.34		
1:24		
1:22		

Item	Allow for protecting thermoplastic tile flr finish

CLAUSE T35

Sheet ①

CLUBHOUSE

TIMBER FLAT ROOF CONSTRUCTION

Construction

```
            7.200
less        0.950
            6.250     Half Roof
            8.100

less  .265  1.235
     1.265  1.765
            6.335
```

100 x 50 mm Down and wallplate   (Carcassing Fir-Larch).
(H Roof)

(L Roof)   *

Bedds 100mm wide wallplate in c.m. (1:3).

```
Low Roof
            8.100
less  .265  6.230
     2/.265 0.530  2.030
                   6.070
less air spaces
to end jsts  0.75 0.150
            ÷.600 ) 5.920
                    9.9
              = 10+1
              = 11 Boards

High Roof
            6.230
less 2/.265 0.530
                   0.100
            ÷.450 ) 5.620
                    12.49
              = 13+1
              = 14 joists
```

```
7.20
6.25
0.95
2/6.34
```

CLAUSE N6 (3) - carcassing timbers automatically assumed to be of "sawn" quality material.

CLAUSE G43 - measured as lineal item stating width of bed.

Calculation for number of roof joists - distance between internal wall surfaces less allowance for air spaces to end joists - resultant value divided by spacing for joists - gives spaces, therefore for number of joists an addition of one is added to this figure.

①

Sheet ②

```
            0.950
            0.265
            ÷.450 ) 1.215
                    2.7
              = 3+1
              = 4 joists
                    Lprofit
            2/.265  1.235
                    0.530
                    1.765
```

63 x 200 mm Down and roof joists   (Carcassing Fir-Larch)

(H Roof)

```
14/3.87
4/1.77
```

"Roofing 12" boards, formed out of 100 x 50mm dwt top batten, flange plus 2mm thick corrugated p/c web glued & fabricated offsite, this includes all offloading & lifting into final position.

(L Roof)

```
joists 7.200
fall 1 in 60.
∴ 7.200 = 120 mm fall
   60
plan size = 10 m
         ∴   130 MAX
              70 MIN
         ∴ Aver ½ ) 140
               70 mm
```

```
11/1.
```

CLAUSE N17 - composite item, fabricated in offsite situation.

CLAUSE N18 - measured as an enumerated item, with full description of composition and allowance for lifting into position.

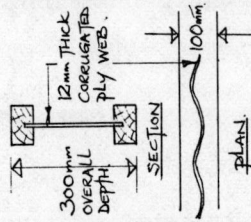

300mm OVERALL DEPTH — 12mm THICK CORRUGATED PLY WEB — 100mm

SECTION

PLAN

②

Panel ③

11/ 7.20	And firrings 50mm # by av. 70mm dp. (L. Roof).
	3.865 / Fall 1:60 = 64.4 mm fall / 60 / plus mile 10.0 / 74.4 MAX / 10.0 MIN / ÷2) 84.4 / ∴ Av # = 42.2 mm
14/ 3.87 4/ 1.77	And firring 50mm # by av 43mm dp. (H Roof).
	less 8.100 / .765 / 7.335 1.765 / 6.345
7.20 / 3.87	19mm # Wrose sheets (Roof grade) 2 T&G edged laid plain firring to fallow fixed to best joist w/ 8No nails. (Measd Net). (H Roof).
2.10 / 0.95	Dett Ditto as last (H Roof).

CLAUSE N6 (1) – firrings measured separately from the roof boarding and given in lineal metres, stating width and average depth.

CLAUSE N4 (1) – classification by situation and measured as superficial item stating method of jointing and fixing to carcassing timbers.

③.

Panel ④

7.20 / 6.35	22mm # Wrose sheets, level of felds a.b.d to top flanges of coupling boards. (L. Roof) (Measd Lap).
7.20 6.35 / 7.20 3.87	75mm # flat glass to BS /785 in solution placed over joists pointing to b/g beds lapped 108mm each side joists
2.10 / 0.95	Dett Ditto as last.

FASCIA:
Mitres 2/.025   7.200   .050
                7.250
                6.345   .025
1 mile   .100
Endlap
6.470
3.865 2.100
1.765
Mitres 2/.025   .050
1.815 2.100
1 mile .025
2.125

CLAUSE N30 – measured superficial to the area covered – with allowance for laps stated in description.

④

Page 5 (left panel):

Calculation for the build up of fascia board depths to opposite side of roof. Fascia board to sides of roof taken as average depth.

	Deep Fascia	Shallow Fascia
H. Roof		
Air dept	·200	·200
fring.l	·065	·010
Bdg.	·019	·019
	0·284	0·229
L. Roof	Deep Fascia	Shallow Fascia
Rafter depth	·325	·325
fring.l	·120	·010
Bdg.	·022	·022
	0·467	0·357

25 mm th. wrot deal T+G bdg as fascia to roof o/a depth 467 mm.
(L Roof) (csa = ·012 m²) | 6·47

Ditto as fascia to roof o/a depth 357 mm.
(L Roof) (csa = ·009 m²) | 6·47

```
0·467
0·357
÷2) 0·824
    0·412
```

CLAUSE N4 (1)g – measured separately in lineal metres stating c.s.a.

⑤

Page 6 (right panel):

Tapered fascia measured stating average depth and raking cutting measured separately as CLAUSE N5(2).

Ditto as last but tapered to av. dep. of 412 mm. dp.
(L Roof) (csa = 0·010 m²) | 7·25

Raking cutting on 25 mm th. fascia bdg. | 6·30 / 0·98

```
3·865
 ·050
3·915
0·750
0·225
0·975
```
Mitred
Mitre

25 mm th. fascia a. bd., o/a depth 230 mm.
(H Roof) (csa = ·006 m²) | 6·30 / 0·98

Ditto as but, b.d.o/a depth 285 mm.
(H Roof) (csa = ·007 m²) | 7·25

```
0·284
0·229
÷2) 0·513
    0·256
```

Ditto as but but tapered to av. dep. 256 mm.
(H Roof) (csa = ·007 m²) | 3·92 / 2·13 / 1·82

Raking cutting on 25 mm th. fascia bdg.

Mitred ends to fascia a. b. d. | 7 / 1

Fitted ends to fascia a. b. d. | 3 / 1

⑥

CLAUSES V1, 2 & 3 - less than 300mm girth, therefore measured separately as lineal item, in 150 mm stages.

CLAUSES M33 & 34 - classification of work according to roof slope, and description of material.

Ditto as last but in work 150-300mm girth

(H Roof)

ROOF COVERING

	7.200
Add Eaves .050	
Fascia .050	.100
	7.300
	6.345
Clad on edge eg	.050
	6.395
	3.865
	.100
	3.965

3 layer bituminous felt roof covering to BS 747, laid to falls, first layer fixed w/ galv. felt nails to fir battens, 2nd & 3rd layers bonded to previous layer, 1st & 2nd layers 1.84 kg/m², refix layer 3.6 kg/m²

Ditto Ditto as last.

Verge	
	.050
Seal bolt	.025
A fillet	.025
proj below bat	0.100
+ drip	0.075
Eaves	
	.050
	.025
	0.075

0.98	
6.30	7.25
2.13	1.82
	3.92
7.30	
6.40	7.30
	3.97
2.10	0.95

CLAUSE N11 (2) - wrot finish as distinct from other rough sawn material, stating cross sectional sizes.

CLAUSE N10 - lineal item stating extreme sizes of member from which fillet is cut. Includes all mitres.

CLAUSES V1, 2 & 3 - classification according to surfaces to which paint is to be applied.

Batts 2/.025	7.250
	.050
	7.300
	6.476
	.050
	6.520
	3.95
	.050
	3.963

Wrot deal batten 25 × 50 mm

deal triangular fillet to verge treated out of 25 × 50 mm

(25 V 50)

Eaves battens	.465	.355	.409
	.050	.050	.050
	.415	.305	.359
Bat Ht.	.015	.025	.025
	.440	.330	.384

k.p.s + 3 f/c general surf as 300 mm girt

(L Roof)

Eaves batt	.284	.259	.265
	.050	.050	.050
	.234	.179	.215
Bat Ht	.025	.025	.025
	.259	.204	.240

3/7.30	
2/6.52	
2/3.97	
2/3.97	
	6.52
2/7.30	
6/.47	0.44
0.38	
6/.47	0.33
7.20	

CLAUSE M56 – Protection of work.

Allow for protecting
all bituminous
felt roofing work
after completion.

	Item

CLAUSE M36 – measured separately
in lineal metres, girth of treatment
stated in description.

2·100
·050
·025
2·175

0·950
·050
·025
1·025

7·200
·950
6·250
·100

2/·050 ·050
3/·025 ·050
6·400

7·30	
6·40	
7·30	
3·97	
2·18	
6·40	
6·40	
1·03	
7·30	

4 Measurement of pitched roofs

As with flat roofs (Chapter 3) the work can be sub-divided: (a) constructional work; (b) roof coverings; (c) rainwater goods.

Calculation of lengths of carcassing timbers. Length of rafter measured is the extreme dimension out of which the member would be cut. Length of rafter may be scaled, if suitable drawings are provided, or may be calculated using either Pythagoras' theorem or trigonometry:

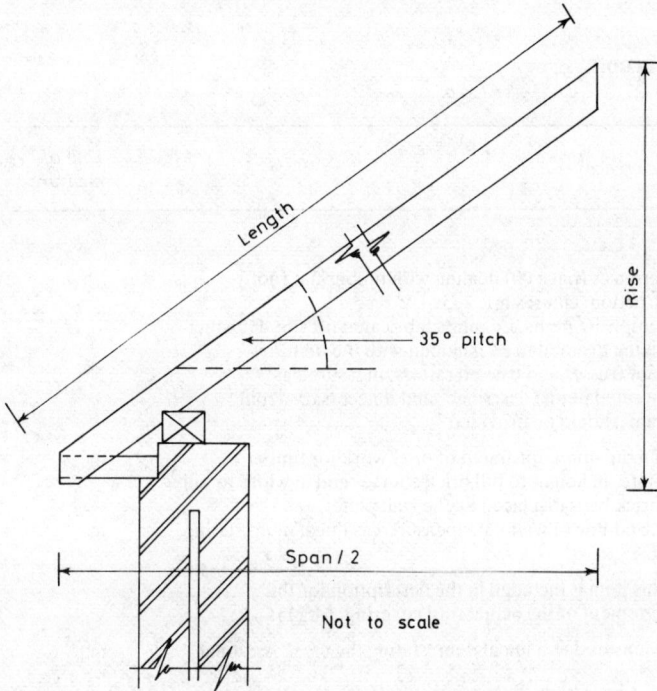

Not to scale

$$\text{Length} = \sqrt{\left[\text{Rise}^2 + \left(\frac{\text{Span}}{2} \right)^2 \right]} \quad \text{or} \quad \frac{\text{Span}}{2} \times \text{secant } 35$$

Hip rafter length can be scaled by drawing out a right-angled triangle, the base (plan length) and vertical height representing rise of roof (see diagram).

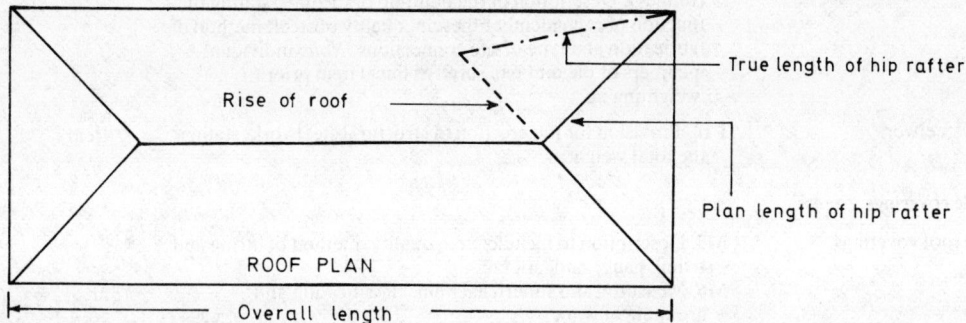

ROOF PLAN

Area of roof coverings is measured by the same method for either a gabled or hipped roof, provided roof pitch is constant for all slopes. Area of roof shown in diagram = 2 (overall length of roof × length of slope).

The following schedule gives a general indication of the common items of work measured in the construction of floor and roofs, and at the same time gives the corresponding clauses of the *Standard Method of Measurement of Building Works*, their implications and the respective units of measurement.

Schedule of SMM requirements for pitched roofs

Item No.	Description of work	SMM reference	Unit of measure
	Timber roof construction		
1	Structural roofing timbers.	See references (page 00) dealing with timber flat roof construction: clauses N1, 2, 3, 4 & 5.	
		N17. Composite items, i.e. units fabricated off site. Hoisting and fixing deemed to be included with the item.	
		N18. Roof trusses and trussed rafters, measured as enumerated items; described, and if necessary a 'bill diagram' should be provided.	No.
2	Adjustment of previously measured brickwork for wallplate.	**G5(2)**. Deduction/adjustment of brickwork for timber wallplate, in height to full brick courses and in width to full half-brick beds displaced by the wall plate.	m^2
		G43(1). Bedding of wallplate measured as lineal item, stating width.	m
3	Battening.	**M16**. This item is included in the description for the measurement of the actual roof covering: M5(1).	
4	Fascias, soffits and barge boarding etc.	**N4(1)**. Measured as a lineal item, stating the cross-sectional area.	m
5	Painting fascias etc.	See references to decoration in section dealing with internal finishes (SMM Section V).	
6	Roof insulation.	**N30**. Measured the net area covered as a superficial item with the allowance for laps etc. given in the description.	m^2
	Steel roof construction		
7	Roof trusses.	**P4**. Generally measured as a lineal item and billed by weight (tonnes). Description of the build-up of the unit, stating the function of each member therein, quality of steel, method of fabrication and type of site connections. *Note:* individual members of the unit measured as lineal item prior to 'weighting up'.	m
8	Erection of steelwork.	**P10**. Provision for the erection of structural steelwork, stating the total weight.	Item
	Slate and tile coverings		
9	Slate or tile roof coverings.	**M5**. Description to include type, quality, method of laying and fixing, gauge and laps etc.	
		M6. Measured as a superficial item, classification and grouping of work.	m^2

Item No.	Description of work	SMM reference	Unit of measure
10	Labours on slate or tile roof coverings.	**M7**. Square, raking and curved cutting measured as a lineal item.	m
11	Eaves, verges, valleys and hip treatments etc.	**M8, 9, 10 & 11**. Measured as lineal items as extra over the basic roof coverings.	m
		N11(3). Mitred angles, intersections on hips etc. measured as enumerated items.	No.
12	Metal slates etc.	**M14**. Usually made by plumber, but fixing by roofer; measured as an enumerated item.	No.
13	Underfelt to roof coverings.	**M17**. Measured net area as fixed, with description including allowances for laps and method of fixing.	m²
	Sheet roofing, i.e. asbestos cement		
14	Sheet roofing generally.	**M18**. Complete description of type and quality of material with laps and method of fixing.	
		M19. Measured as a superficial item, classification and grouping of work.	m²
15	Labours on sheet roof coverings.	**M20**. Square raking and curved cutting, measured as a lineal item.	m
16	Filler pieces, ridges, hips, barge boards etc.	**M21, 22, 23 & 24**. Measured as lineal items.	m
17	Flashings etc.	**M25**. Flashings and expansion joints etc. measured separately as lineal items.	m
18	Protection.	**M56**. given as an item.	Item
	Above ground drainage (rainwater goods)		
19	Gutter work.	**R7(1)**. Measured as a lineal item, with the size and type being included in the description. Joints generally deemed to be included with the lineal measurement of the gutter.	m
20	Gutter fittings and supports.	**R7(3)**. Fittings, i.e. bends, stopped ends etc., enumerated as extra over the gutter.	No.
		R8(1). Standard supports to be included in the description of the gutter work.	

Sequence of measurement for pitched roof

Timber construction

1. Wallplates, bedding and adjustment of brickwork.
2. Rafters and collars.
3. Ceiling joists.
4. Hip rafters.
5. Ridge board.
6. Insulating quilt.

Tile covering

7. General tiling to roof slopes and felt underlay.
8. Double course at eaves.
9. Half-round ridge capping.
10. Half-round hip capping.
11. Raking cutting on tiles and felt.

57

12. Intersections between ridge and hip capping.
13. Filling ends of hip tiles and hoop irons.

Eaves treatment

14. Fascia board.
15. Soffit board.
16. Painting to items 14 and 15.
17. Rainwater gutter with fittings measured extra over.
18. Rainwater pipe with fittings measured exra over.

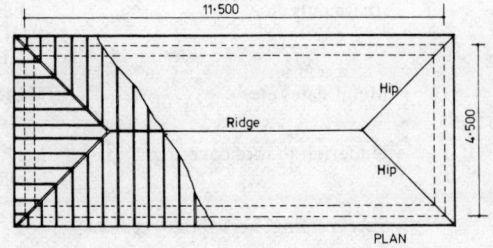

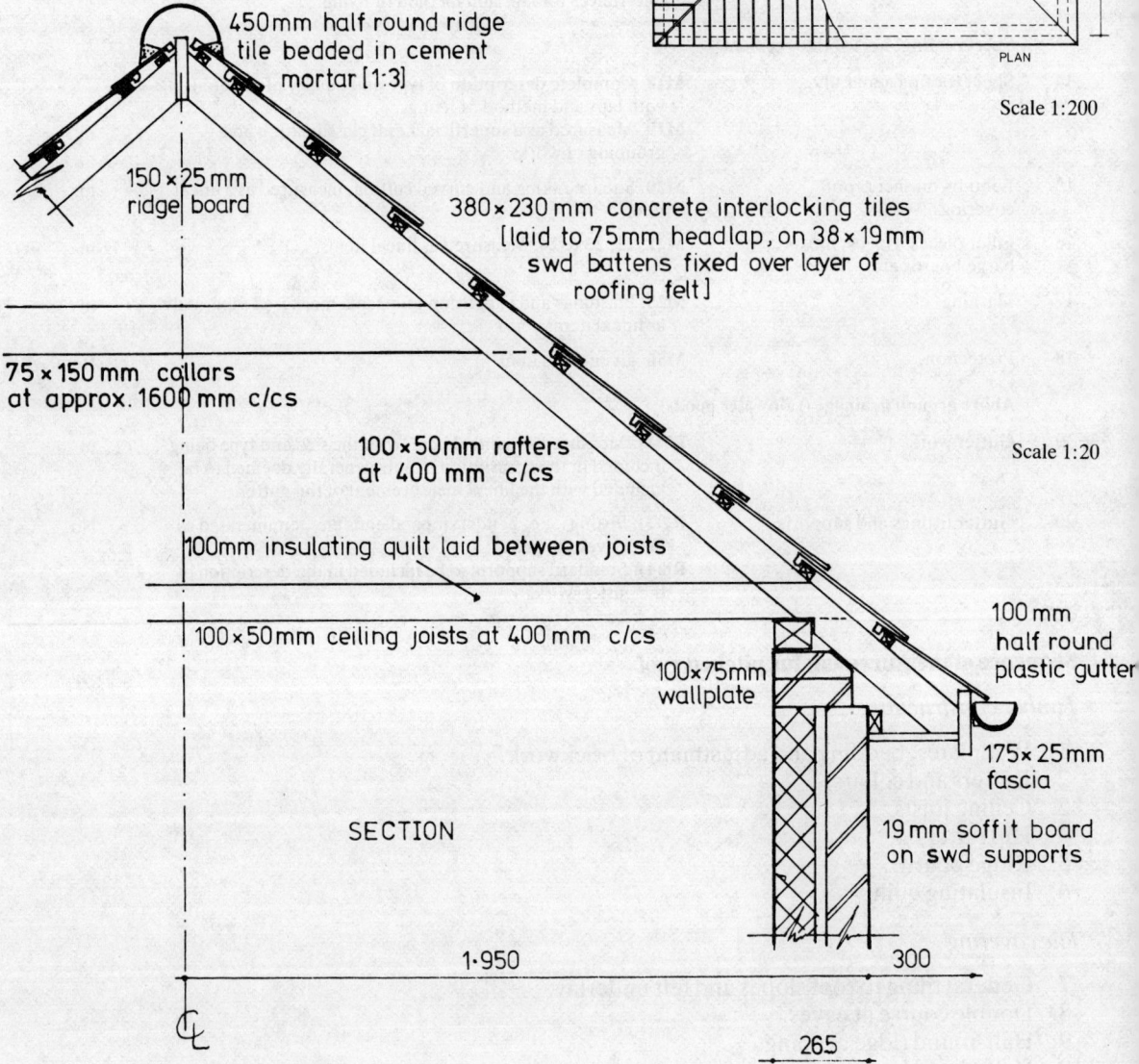

PLAN

Scale 1:200

450mm half round ridge
tile bedded in cement
mortar [1:3]

150 × 25 mm
ridge board

380 × 230 mm concrete interlocking tiles
[laid to 75mm headlap, on 38 × 19 mm
swd battens fixed over layer of
roofing felt]

75 × 150 mm collars
at approx. 1600 mm c/cs

100 × 50mm rafters
at 400 mm c/cs

Scale 1:20

100mm insulating quilt laid between joists

100 × 50mm ceiling joists at 400 mm c/cs

100 × 75mm
wallplate

100 mm
half round
plastic gutter

175 × 25 mm
fascia

19 mm soffit board
on swd supports

SECTION

1·950

300

265

58

PITCHED ROOF

Construction

w/plate
11.500
4.500
×2)6.000
32.000
Rvd Corned 4/21.365 2/120
Add faces 8   29.800
Add wip 8
Cled plte
Corner 4/2.00   0.800
30.680

100 × 50 mm sawn
duct softplate
(Carcassing timber)

Add corner 4/100 .400
30.280

Barby wallplate
100mm wide in c.m.
(1:3)

Add corner 4/215 .860
30.740

Datt
1B wall in cement
in 9m (1:1:6) faced
in Eng bd bond
(EAVES FILLING)

		30.68
		30.28
		30.74
		0.08

(1)

CLAUSE G43 - measured as lineal item - width of bed must be stated.

Add corner 30.740
4/4/102b 0.205
30.945

Ddt
1B wall in cement
in 9m (1:1:6) faced
in street bond
(EAVES FILLING)

Roof plating 11.500
.600
2/300 12.100
÷400 12.100
30.25
= 31 + 1
= 32 Rafters

Plus extra Rafter
@ ck of hipend
end. 2/1 = 2N°

Rafter-to-wipt
SCALED = 2.900

50 × 100 mm sawn duct Rafter in
pitched roof
(Carcassing timber)

| 2/32/2.90 | | |
| 2/1/2.90 | | |

Post 11.500
2nd .600
w.wall 2/165 .530
Arch-ansd 2/4025 .050 1.180
÷400)0.320
25.80
= 26 + 1
= 27 Joists
2/1.900 = 3.800

(2)

CLAUSE N2 (1) - classified in various groups of work as stated.

		Less 2/265 11·500
		0·530
		10·970
		4·500
		·530
		3·970
		10·970 ·200
	11·17	11·170
	4·17	3·970
		·200
		4·170

100 m.m. insulating quilt to BS 5785 (fibre glass) laid between c.g. joists at 100 mm laps.

CLAUSE N30 – measured superficial to the area covered – with allowance for laps stated in description.

COVERINGS.

o/Meas 2/3·00   11·500
·600
12·100

2/12·10	380 x 230 mm interlocking
2·90	conc. tiles laid to 75 mm lap, on 38 x 19 mm sawn swd battens (cond. tile battens) + all tiles nailed – every alt. course.

CLAUSE M5 – full description of materials and sizes used.

CLAUSE M17 (1) – measured superficial – stating the laps.

7·00	Lay'g of special ditto 2 rett. to BS 473, lapped 150 mm at r'ge & nailed to swd batten at 25 mm galv. c.l. nails

CLAUSE M11 (3) – measured as a lineal item.

&c. a. roof tilg. for h.v. ridge tile lapp'g bedd'd + ptd. in c.m. (1:3).

27/3·80	Sawn swd ceiling joists 50 x 100 mm. (carcassing timber)

$\dfrac{\text{No. of colls.}}{32/4} = 8 \text{ N}^o.$

2/1·040
2·080

NB. It has been assumed for this exercise that the internal wall layout is such that it provides the required degree of intermediate support for the ceiling joists; otherwise it would be necessary to measure at this stage items for sawn softwood in "Binders and Hangers".

8/2·08	Sawn swd in roof rafters 75 x 150 mm (carcassing timber)

Calculation for actual length of hip rafter based upon height of roof and plan length of hip – scaled from drawing.

$$Hip = \sqrt{1\cdot9^2 + 3\cdot6^2}$$
$$= \sqrt{16\cdot57}$$
$$\therefore Hip = 4\cdot070$$

4/4·07	Sawn swd hip rafter 50 x 200, incldg. cutt'g + fitg. ends of rafters (carcassing timber)

Ridge long't
SCALED = 7·000

7·00	Sawn swd ridge board 25 x 150 mm incldg. cutt'g + fitg. ends (length 6·976) (7·2nd long)

(6)

Allow for protecting roof tile covering. — **Item**

```
                    11·500
                     4·500
                 ×2)16·000
                    32·000
less fascia   ·300
              ·025
              ·275
          2/·275  ·550
                 11·500
                 12·050
                  4·500
                   ·550
                  5·050
                 12·050
              ×2)17·100
                 34·200
```

EAVES TREATMENT

19mm ← cont. board eaves bdg 275mm wide. — **34·20**

CLAUSE M56 – protection of work.

CLAUSE N4(1)g – lineal item, stating cross sectional area.

Raking cutting on board. — **4/1**

Roof down and eaves board plugg'd to brick. (25×50 mm) (c.s.a = ·001/m²) — **32·00**

Add fascia 4/2/·025 0·200
                      34·200
                      34·400

CLAUSE N1 (10)² – cross sectional area exceeds 0.002m² therefore raking cutting (mitres) enumerated.

(5)

```
                 11·500
                  4·500
              ×2)16·000
                 32·000
Add'd erned 4/2/·300  2·400
                 34·400
```

E.o. roof tiling for double course at eaves, undercourse of 267×165 plain roof tiles. — **34·40**

E.o. roof tiling for h.r hip capping bedd + ptd in c.m (1:3). — **4/4·07**

CLAUSE M11 (3) – measured in same manner as ridge capping.

Cutting roof tiling to hips.
&
Raking cutting in bituminous felt. — **2/4/4·07**

CLAUSE M11 (1) – measured as lineal and to both sides of hip line.

CLAUSE M17 (2) – lineal item for cutting on underlay.

Mitred intersection between ridge + hip capping. — **2/1**

Ends of hip tiling bedd in c.m (1:3) + the slips
&
Seal top in c.m. bedd to feet of hand hip tiles. — **4/1**

CLAUSE M13 – enumerated item giving method of fixing.

CLAUSE R7 (3) - fittings measured extra over the basic gutter work.

E.o. gutter for extl angles.

E.o. gutter for outlet w. nozzle for a 75 mm ⌀ downpipe.

75 mm ⌀ plastic r.w. pipe w. socketed jt incl jts in running lengths.

CLAUSES R9 & 10 - measured as lineal item over all fittings which are later enumerated as extra over.

E.o. r.w. pipe for swanneck 300 mm projection.

&

E.o. ditto for r.w. shoe.

Plastic pipe clips for 75 mm ⌀ pipe plugged + screwed to brick.

4/	1
2/	1
2/	3·50
2/	1
2/	3

CLAUSE N4 (1)g - measured separately in lineal metres.

25 mm t.wrot deal fascia bdg incldg all mitres (175mm dp.)
(c.s.a = ·004m²)

```
      275
  40
175 |_____
 25
```

·175
·025
·040
0·240
34·200
1·100
33·100

less 4/·275

Nailed to fascia bdg.

CLAUSE V3 - measured as a lineal item not exceeding 300 mm girth.

k.p.s. + ③ to general surfs 150-300 mm girth (FASCIA)
Extly. (SOFFIT).

CLAUSES R6, 7 & 8 inclusive - full description to include method of jointing and supporting gutters.

RAINWATER GOODS

Fascia 34·400
Gutd 4/no ·400
34·800

100 mm ⌀ plastic h.v eaves gutter to BS.4576 fixd w. gutter straps w. clips & jtd in plastic brackets screwd to deal fascia incldg jts in running lengths.

	34·40
	34·40
	0·24
	33·10
	0·28
4/	1
	34·80

5 Measurement of framed buildings

Covered by the rules of measurement laid down in the *Standard Method of Measurement of Building Works*, sections F and P. A wide variety of work is covered in this aspect of a building and can be broken down and measured under the following headings:

- Concrete frames: (a) in situ concrete, (b) precast concrete, (c) prestressed concrete work.
- Steel-framed structures.

Concrete frames

All sections of concrete work incorporate three aspects: (a) formwork, (b) reinforcement, and (c) concrete. The formwork and concrete aspects frequently occur in steel-framed structures owing to the need for the provision of a fire-resisting encasement to the steel elements.

The general sequence of measurement for concrete structures follows the order of:

1. Concrete, reinforcement and formwork to columns.
2. Concrete, reinforcement and formwork to beams: (a) main beams, (b) secondary beams.
3. concrete, reinforcement and formwork to slabs, including the necessary adjustments for work relating to beams (T or L beams) cast as an integral part of a floor or roof slab. There is a separate category of measurement for concrete to 'deep beams'.

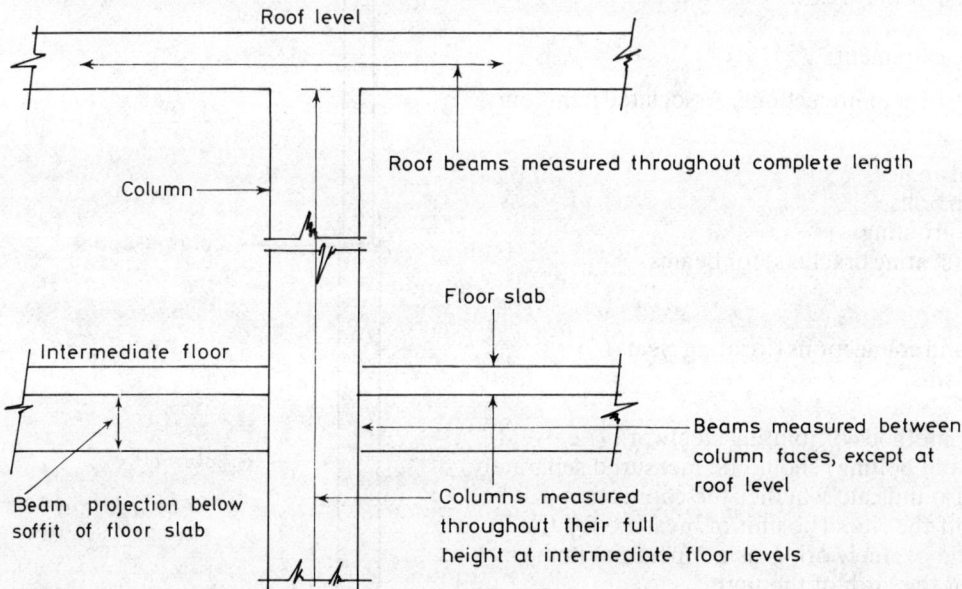

Two main types of formwork in framed structures: (a) formwork to sides and soffits of beams, one overall item irrespective of cross-sectional shape, i.e.

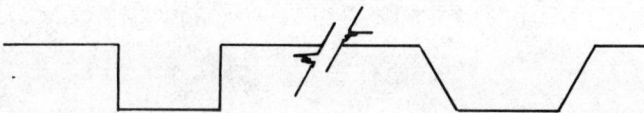

(b) formwork to vertical or battering sides of columns, grouped irrespective of cross-sectional shape of column (measured the net area in contact with the concrete). These two types are measured separately and described as not contiguous with another member.

The measurement of steel reinforcement in concrete items is frequently simplified by the use of 'Bar Bending Schedules'. These specify: structural member, type of bar, size, shape, length, number of bars.

Steel frames

Measured as either:

- unfabricated steelwork cut to length only (as steel in an unframed building), or
- fabricated steelwork, e.g. shop fabrication and connection of cleats and brackets etc. (as steelwork for a framed structure).

Sequence of measurement and groupings:
1. Stanchions and associated work.
2. Main beams and connections.
3. Secondary beams and connections.
4. Struts and braces.
5. Purlins and rails.
6. Steelwork in roof trusses.

Method of measurement:

Overall length of column sections. Associated items on column:

Baseplate and angles.
Holding down bolts.
Wedging and grouting.
Connections (seating brackets) for beams.
splice plates.
Cappings.
Main beams and connections (locating cleats).
Secondary beams.

The different methods of jointing steelwork (i.e. welding, riveting and bolting) should be measured separately, and should also indicate whether the connection will be made on or off the site. The unit of measure for the offsite painting of steelwork is as a superficial item (m^2), irrespective of the girth of the unit.

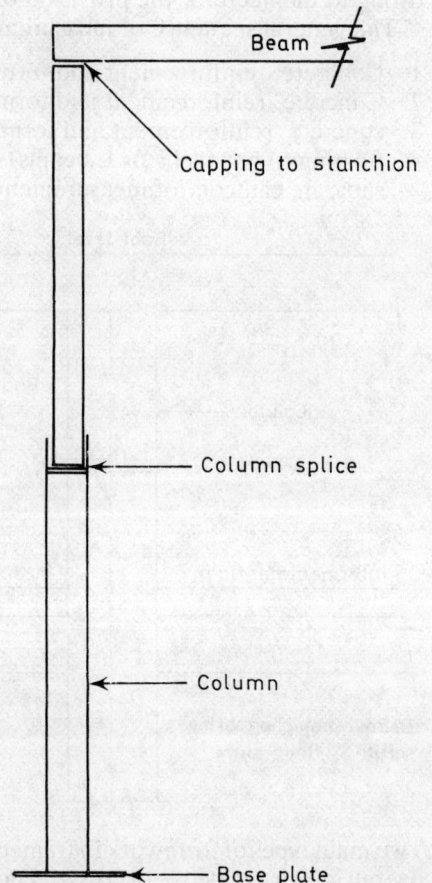

Labels in diagram: Beam, Capping to stanchion, Column splice, Column, Base plate

The following schedule gives a general indication of the common items of work measured in the construction of concrete or steel framed structures, and at the same time gives the corresponding clauses of the *Standard Method of Measurement of Building Works*, their implications and the respective units of measurement.

Schedule of SMM requirements for framed buildings

Item No.	Description of work	SMM reference	Unit of measure
	Concrete work		
1	Concrete generally.	**F4(1)**. Descriptions to include type of materials and mix proportions. **F5(1)**. Size classification: where concrete is placed in members of small cross-sectional area such as beams and columns, the work should be described as in one of the following groups: not exceeding 0.03 m^2; 0.03–0.10 m^2; 0.10–0.25 m^2; exceeding 0.25 m^2. **F4(2)**. Reinforced concrete work described separately.	
2	Concrete in floor or roof slabs.	**F6(9)**. Suspended slabs are measured over the bearings. Measured as a cubic item but stating the thickness in compliance with clause F5(2): not exceeding 100 mm; 100–150 mm; 150–300 mm; exceeding 300 mm.	m^3
3	Concrete in beams and columns.	**F6(9, 14 & 15)**. Measured as a cubic item in separate groups, in accordance with the cross-sectional area requirements in F5(1).	m^3
4	Finishes to concrete.	**F8**. Finishes such as mosaic, granolithic etc. cast on to concrete – measured superficially as an extra over item, stating the mix and thickness.	m^2
	Steel reinforcement		
5	Reinforcement generally.	**F11 & 12**. Description should include the quality, size and weight of fabric reinforcement. For bar reinforcement description should include the quality, size, bends, hooks and tying wire etc.	
6	Bar reinforcement.	**F11**. Classification by situation of usage, and separation into straight bars, bent bars, curved bars and links etc. *Note:* this is an exceptional case, being measured as a lineal item, but billed by weight (tonnes). (Exceptionally long bars measured separately: horizontal bars over 12.00 m in further 3.00 m stages; vertical bars over 5.00 m in further 3.00 m stages.)	m
7	Fabric reinforcement.	**F12**. Measured net as the area covered as a superficial item, with no allowance for laps.	m^2
	Formwork for concrete		
8	Formwork for concrete.	**F13**. Measured as a superficial item to the actual concrete faces being supported, distinguishing between the type of surface finish required to the concrete. *Note:* description is deemed to include for easing, striking and removal of formwork.	m^2
9	Formwork to surfaces.	**F13(i)**. Classification according to situation of usage and positional limitations, e.g. formwork required at excessive heights or isolated situations.	

65

Item No.	Description of work	SMM reference	Unit of measure
10	Formwork to edges of slabs and upstands generally.	**F15(8)**. Measured as a lineal item, stating the depth in the descriptions as: not exceeding 250 mm; 250–500 mm; exceeding 500 mm.	m
11	Protection.	**F45**. Given as an item.	Item
	Structural steelwork		
12	Structural steelwork generally.	Generally measured in metres, but billed by weight (tonnes). **P3**. work grouped under the headings of: (i) *unfabricated steelwork*, i.e. odd items of steelwork occurring in a building of traditional construction; (ii) *fabricated steelwork*, i.e. a skeletal framework of beams and columns as used in a framed structure. **P5**. Designation of members. **P7**. Steelwork measured net as executed, with no adjustments for item such as holes, slots etc. Also measured as overall lengths required to form the members (no deductions for splays etc.).	
13	Steel beams, stanchions, braces and ties etc.	**P4(3)**. Descriptions to include the structural function of the member and its position in the work. Built-up units to list the components in their formation. Proprietary members (castellated, litzka beams etc.) should be described separately.	
14	Fittings.	**P8(1)**. Measured separately following the measurement of the item upon which they occur: e.g. baseplates, splice plates, cappings, brackets and cleats etc., measured superficial or lineal as appropriate and then weighted up. **P8(2)**. Bolting down and wedging for stanchions (disregarding the number of wedges used). Site welded connections etc. measured separately as enumerated items.	No.
15	Painting steelwork.	**P9**. Offsite painting measured superficial irrespective of the girth of the member, *or* site painting measured in compliance with SMM Section V.	m^2
16	Erection of steelwork.	**P10**. Provision should be made for the operations involved in the erection of the steelwork within the building, stating the total weight.	Item
17	Protection.	**P11**. Given as an item.	Item

Sequence of measurement for concrete work in link corridor above level of ground floor slab

1. Concrete in roof slab.
2. Concrete in beams (attached shallow beam therefore still measured as concrete in roof slab).
3. Concrete in columns.
4. Formwork to soffits of roof slab.
5. Formwork to edges of roof slab.
6. Deduct formwork to soffit of roof slab.
7. Formwork to sides and soffits of beams.
8. Formwork to vertical sides of columns.
9. Reinforcement to roof slab.
10. Reinforcement to roof beams.
11. Reinforcement to columns.

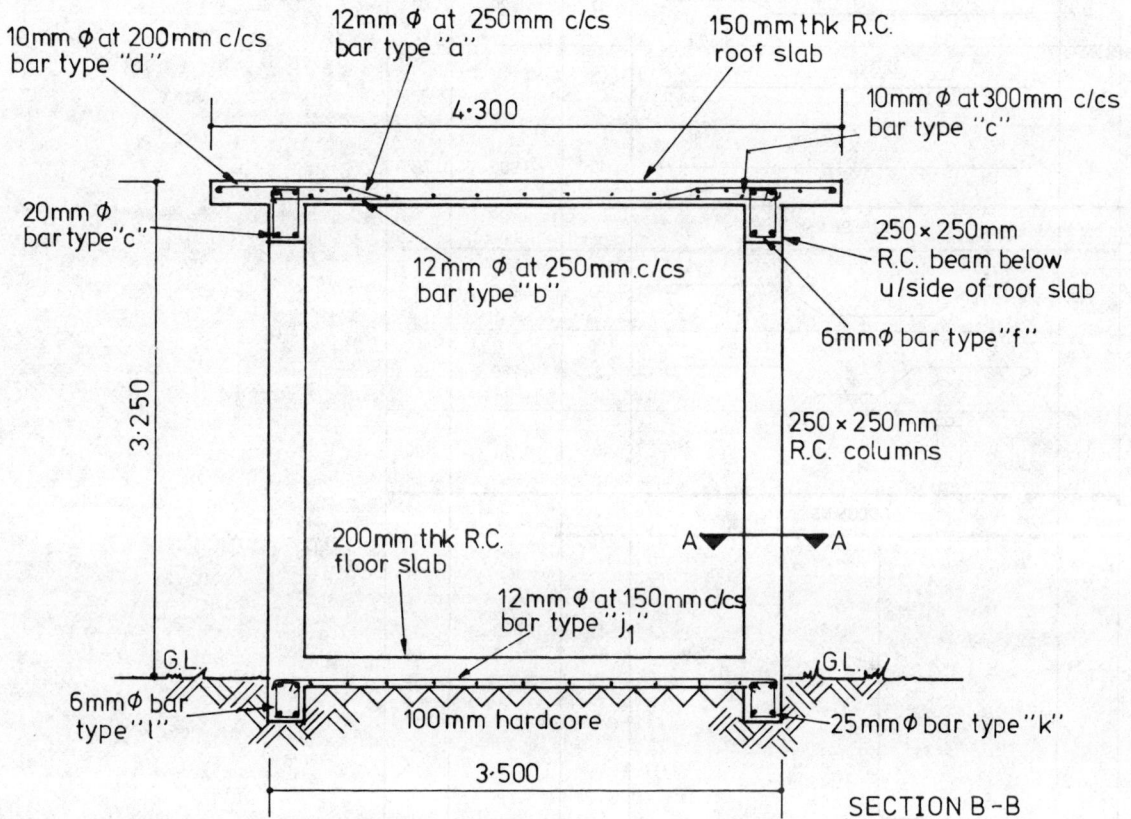

10mm ⌀ at 200mm c/cs
bar type "d"

12mm ⌀ at 250mm c/cs
bar type "a"

150mm thk R.C.
roof slab

10mm ⌀ at 300mm c/cs
bar type "c"

4·300

20mm ⌀
bar type"c"

12mm ⌀ at 250mm.c/cs
bar type"b"

250 × 250mm
R.C. beam below
u/side of roof slab

6mm⌀ bar type"f"

250 × 250mm
R.C. columns

3·250

A ▼   ▼ A

200mm thk R.C.
floor slab

12mm ⌀ at 150mm c/cs
bar type "j₁"

G.L.

G.L.

6mm⌀ bar
type "l"

100mm hardcore

25mm⌀ bar type"k"

3·500

SECTION B-B

Scale 1:50

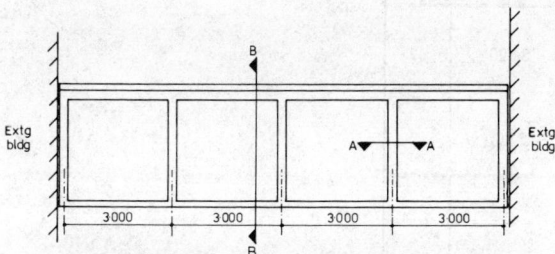

Extg
bldg

B

A ▼   ▼ A

Extg
bldg

3000   3000   3000   3000

B

ELEVATION OF REINFORCED CONCRETE FRAMEWORK TO PROPOSED
NEW LINK CORRIDOR TO BE CONSTRUCTED BETWEEN EXISTING BUILDINGS

Scale 1:200

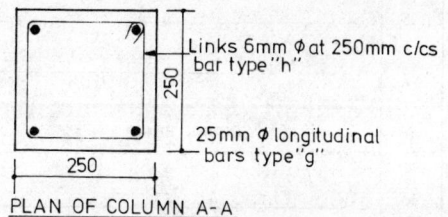

Links 6mm ⌀ at 250mm c/cs
bar type "h"

250

25mm ⌀ longitudinal
bars type"g"

250

PLAN OF COLUMN A-A

Scale 1:100

67

BAR BENDING SCHEDULE				
BAR	PROFILE	DIAM. mm	OVERALL LENGTH	N°. OFF
ROOF SLAB				
a.		12	4·220	49
b.		12	4·220	49
c.		10	3·900	48
d.		10	3·900	48
ROOF BEAMS				
e_1		20	4·080	8
e_2		20	4·780	8
f.		6	1·180	120
COLUMNS				
g.		25	3·700	40
h.		6	760	120
FLOOR SLAB				
j_1		12	3·620	82
j_2		10	3·900	48
GROUND BEAM				
k.		25	7·430	16
l.		6	840	160

Column ① (left)

LINK CORRIDOR

REINFORCED CONCRETE WORK
ABOVE FINISHED LEVEL OF THE
GROUND FLOOR SLABS ONLY

Concrete

Slab Lengt
2/3,000 12.000
2/3/250   0.250
          12.250

1/2.25	4.30
	0.25
	0.15

R. conc (1:2:4/20mm to Cap) 100–150mm to tampd and reinfd to suspd horiz roof slab.

> CLAUSE F6 (9) – measured as a cubic item, over all bearings – stating thickness as CLAUSE F5 (2).

Beam Lengt
3.000
2/4/250 0.250
         2.750

Rod 2/4/250 ...

2/4/	2.75
	0.25
	0.25

Ditto in suspd slab 100–150mm tampd and reinfd to frame.

> CLAUSE F6 (9) – cubic item, additional thickness of concrete below soffite of suspended slab classified in relation to thickness for slabs in CLAUSE F5 (2).

2/5/	2.75
	0.25
	0.25

Ditto in sloped slab 250mm avg on face tampd and reinfd to frame. (C.S.a 0·3–0·6 m²)

> CLAUSE F6 (15) measured separately stating cross-sectional area as CLAUSE F5 (1). (Height scaled between top of G.F. slab and underside of roof beam).

Formwork

Smk to u/s of horiz suspd conc roof slab, suspd aft struck. (* No. dif surface)

| 1/2.25 | 4.30 |

> CLAUSE F15 (1) – superficial item with classification into various groupings according to situation.
> CLAUSE F13 (9) – quality of finish to concrete should be given. NB. Easiest way of doing this is still to distinguish between WROT and SAWN qualities of timber required to provide either a SMOOTH or ROUGH finished surf.
> • (1 No. separate area only due to later adjustment for separate measure of formwork to projecting eaves)

①

Column ② (right)

Dett Ditto as last

| 2/4 | 2·75 |
| | 0·25 |

Smk to face of attached horiz beams, suspd aft struck. (2/2 No. Members)

| 2/ | 2·25 |

> CLAUSE F15 (5) – measured as a lineal item, stating the number of members.

Smk to supporting canted rf horiz roof slab, suspd fin as a slab depth 150mm. (2/2 No. Members)

| 2/ | 2·25 |

> CLAUSE F15 (5) – measured separately as a lineal item stating width in description.

Dett Smk to u/s of horiz suspd conc roof slab a.b.l.

| 2/ | 2·25 |
| | 0·40 |

Smk to vert sides of conc slab, suspd aft struck.

2/3/	2·75
	0·25
2/2/3/	2·75
	0·25

> NOTE – Adjustment for face of end columns, which must be cast against the face of the existing brickwork surfaces.

②

CLAUSE F45 – protection of all reinforcement and concrete work.

6 mm ⌀ Ditto and do. (Bars Mark h.).

Allow for protecting all conc. work after completion.

120/0.76

CLAUSE F17 (a) – full description of material.
F17 (b) measured as executed and deemed to include all tying wire. Given in tonnes.

NOTE:- All dimensions and quantities of reinforcing bars are as extracted from the bar bending schedule.

REINFORCEMENT

Roof Slab

12 mm ⌀ m.s. reinf bars to BS 4449 in suspd. roof slab. (Bars Mark a-b).

10 mm ⌀ Ditto and do. (Bars Mark c-d).

Roof Beams

20 mm ⌀ m.s. Reinf bars a.b.d in roof beams (Bars Mark e_1 v e_2).

6 mm ⌀ Ditto and do. (Bars Mark f).

R.C. Columns

25 mm ⌀ m.s. reinf bars a.b.d. b.d. in conc. cols. (Bars Mark g).

2/49/4.22

2/48/3.90

8/4.08
8/4.78

120/1.18

40/3.70

6 Measurement of internal finishes, doors and windows

Covered by rules of measurement laid down in the *Standard Method of Measurement of Building Works*, sections G, K, N, Q, T and V. These three topics are grouped together in one section as their actual measurement is usually carried out from previously prepared schedules, which are either supplied by the architect to facilitate the taking-off process or alternatively made up by the contractor to assist with the ordering of materials etc. The main aim in each case is to collect and tabulate the information in an easily readable form.

Internal finishes

These are measured as if there were a complete absence of doors and windows in the building, with the necessary adjustments for over-measure being made when the doors/windows themselves are measured.

To ease the taking-off of the quantities of work a logical sequence must be followed, i.e. working through the building floor by floor and room by room, except that where possible similar items should

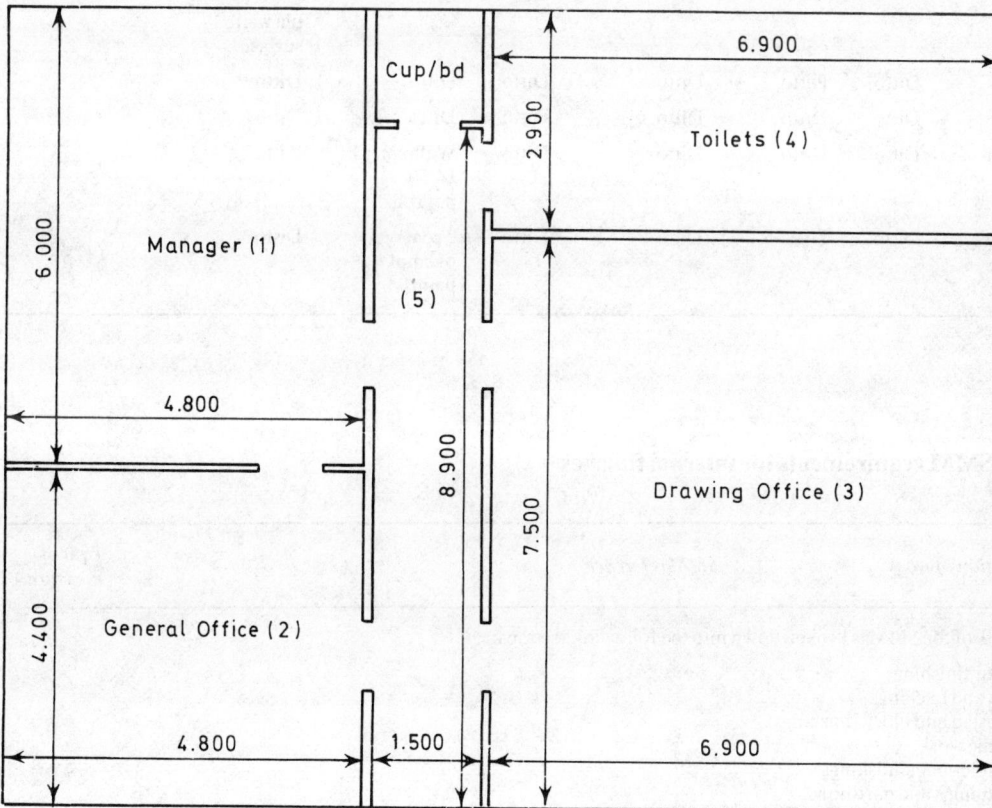

Floor to ceiling height = 3.000 m

Scale 1:100

be gathered together to reduce the overall work load. To eliminate the possibility of errors, each item when measured should be clearly cancelled on the schedule.

The work is mainly superficial; ceilings, walls and floor finishes are measured in square metres, followed by any linear items – skirtings, cutting in painted dados etc. Measurements are taken on the actual finished faces of the walls, floors and ceilings.

Internal finishes schedule

| Room no. | Ceilings | | Floors | | Walls | | Skirtings | Remarks |
	Applied finish	Decor	Non-struct	Applied finish	Applied finish	Decor		
1	Pl'bd & skim	2 coats of emul paint	None	Hardwood strip floor on battens fixed to conc	R & S on bwk	Wallpaper PC £4.50 per roll	100 × 25 mm swd splayed skirting plugged to bwk	
2	Ditto	Ditto	40 mm s/cement screed	Thermoplastic tiles	Ditto	2 coats of emul paint	75 mm high plastic skirting glued to pla wall surface	
3	Ditto	Ditto	Ditto	Ditto	Ditto	Ditto	Ditto	
4	Ditto	Ditto	Ditto	Ditto	Ditto	Ditto	Ditto	
5	12 mm insulation board	Ditto	Ditto	Ditto	Ditto	Wallpaper PC £4.50 per roll	Ditto	
6	Ditto	Ditto	Ditto	Ditto	Ditto	2 coats of emul paint	Ditto	

Schedule of SMM requirements for internal finishes

Item No.	Description of work	SMM reference	Unit of measure

Section T of the SMM is broken down into the following sections:

1. In situ finishings.
2. Beds and backings.
3. Tile, slab and block finishings.
4. Mosaic work.
5. Flexible sheet finishings.
6. Dry linings and partitions.
7. Suspended sheet linings (ceilings).
8. fibrous plaster work.
9. Fitted carpeting.

Item No.	Description of work	SMM reference	Unit of measure
	In situ finishings		
1	In situ finishes generally.	**T3(1)**. Classification of work into internal work and external work. **T3(2, 9 & 10)**. Limitations of work situation or quantity of work involved. **T4**. Description should give a complete outline of the materials, composition, mix, number of coats and the background to which they are applied.	
2	Plasterwork to walls and ceilings.	**T5(1)**. Wall and ceiling finishes measured as separate superficial items. **T4(2)**. Where work does not exceed 300 mm in width this should be measured separately and described. *Note:* measured as the area in contact with the backing surface.	m²
3	Angles on plasterwork.	**T5(5 & 6)**. External angles, rounded internal and external angles, measured as lineal item.	m
4	Labours on plasterwork.	**T11**. Fair joints etc. measured as lineal item to cover the extra labour involved.	m
5	Metal angle beads.	**T5(7)**. Measured as a lineal item with a description of type, size and method of fixing.	m
6	Plasterboard to walls and ceilings.	**T16(1)**. The description should include type of material, background and method of fixing. **T17(1)**. Wall and ceiling finishes measured as separate superficial items in compliance with T5, which deals with in situ plasterwork.	m²
7	Labours on plasterboard.	**T11**. As for in situ finishes.	m
	Decoration work		
8	Decorative work generally.	**V1(2)**. Classification into: (i) new work or redecoration, (ii) internal or external work.	
9	Painting on general surfaces.	**V3(1)**. Description should give an outline of the materials to be used, backing and number of coats. Measured as a superficial item where width exceeds 300 mm wide and as a lineal item when not exceeding 300 mm wide, stating girth in 150 mm stages. **V4**. Classification of work into groupings in relation to the surface to which the treatment is applied.	m² m
10	Wallpapering to walls and ceilings.	**V12(1)**. Description to outline type and quality of material and backing. Measured separately as superficial items. **V12(2)**. Classification of work in relation to surface to which treatment is applied.	m²
11	Protection of decoration.	**V13**. Given as an item.	Item
	Floor finishes		
12	In situ floor finishes.	See previous references in this schedule to T3 and T4. **T8**. Work classified as to whether it is laid (i) level, to falls and/or to crossfalls; or (ii) horizontal, or exceeding 15 degrees to the horizontal. Measured as a superficial item, but where work does not exceed 300 mm wide it is measured separately.	m²
13	In situ skirtings and kerbs.	**T10**. Measured as a lineal item with height/width included in the description.	m

Item No.	Description of work	SMM reference	Unit of measure
14	Tile floor finishes.	**T14**. Full description of type, quality, bedding and jointing materials. Measured as a superficial item in compliance with T4 to T12.	m^2
15	Labour on floor finishes.	**T11**. As for plasterwork.	m
16	Floor screeds.	**T13**. Measured as a superficial item in compliance with T4 to T10.	
17	Protection of floor, walling and ceiling finishes.	**T35**. Given as an item.	Item

Sequence of measurement for internal finishes

Ceiling finishes

1. Plasterboard and skim to ceiling and two coats of emulsion paint – rooms 1, 2, 3 and 4.
2. Insulation board finish and two coats of emulsion paint – room 5.

Wall finishes

3. Render and set treatment to wall surfaces and two coats of emulsion paint – rooms 2, 3 and 4.
4. Render and set on walls a.b.d. and wallpaper finish – rooms 1 and 5.

Floor finishes

5. Hardwood strip flooring to sub/concrete floor – room 1.
6. Sand/cement to receive thermoplastic tiles and thermoplastic floor tiles measured separately – rooms 2, 3, 4 and 5.

Doors and windows

These items are again measured from schedules incorporating the necessary adjustments for the previous over-measure of brickwork and finishes.

Doors. As a general approach, work is measured in the following groups:

1. Door plus associated work (decoration, glazing, ironmongery).
2. Door frame or linings plus associated work (decoration).
3. Adjustments for opening (brickwork or blockwork, wall finishes, finishes to reveals of openings, treatment to head and bottom of opening).

Windows. Work grouped and measured as:

1. Window and all associated items (glazing and decoration).
2. Adjustments for opening (as for doors).

INTERNAL FINISHES:

CEILINGS

6.00	4.80
4.80	4.40
7.50	6.90
6.90	2.90
8.90	1.50

Dim
2d

1) 9.5mm ± gypsum plasterbd to BS 1230 fixd to sawn clg joists w/ 32mm galv clout nails, scrim to all jts and wallpaper junctions.
&

2)

3) 3mm ± skim ct of gypsum pls to BS 1191, ct pls/bd clg.
&
Prep and 2 cts of emulsion pt to pls clgs.

4)

12mm ± insulation bd nailed to stud joists w/ galv clout nails
&
Punch and stop all nail heads and 2 cts of emulsion pt on insulation bd clg.

5)

CLAUSE T16 - quality and type of materials to be used. Work to walls and ceilings given separately as CLAUSE T17.

CLAUSE T4 - skim coat of plaster measured separately as an insitu finish.

CLAUSE V4 - classified as work on ceilings, with description as CLAUSE V3.

CLAUSE T3 (2). small areas of work (in compartments not exceeding 4.00m² floor area) must be measured separately.

1.50	1.50
2/4.80	3.00
2/4.40	3.00
2/7.50	3.00
2/6.90	3.00
2/6.90	3.00
2/2.90	3.00
2/1.50	3.00
2/1.50	3.00

6) 12mm ± insulation bd a.b.d., but in compartments n.ex 4.00m² on plan.
&

Punch and stop all steel heads & 2 cts of emulsion pt a.b.d., but in compartments n.ex 4.00m² on plan.

1) 10mm ± render in cement browning and 3mm ± set render to BS 1191 on brk & blk walls.
&

2)

3) Prep + 2 cts of emulsion pt on pls walls.

4)

5) 10mm ± render + 3mm ± fin to BS 1191 a.b.d, but in compartments n.ex 4.00m² on plan.
&

6) Prep + 2 cts of emulsion pt on pls walls but in compartments n.ex 4.00m² on plan.

WALLS

CLAUSES T4 & 5 - full wall height measured ignoring deductions for skirting grounds where appropriate.

CLAUSE T3 (2)-working compartments not exceeding 4.00m² on plan.

CLAUSE T14 – tile or proprietary skirtings measured as lineal item with extreme height stated as CLAUSE T1...

CLAUSE V13 – allow for protecting work.

CLAUSE N4 – no allowance for joints.

2/4.80	75 mm high plastic skirting glued to pva
2/4.40	2) neutral surface or projecting adhesive to backing all mitres
2/7.50	3)
2/6.90	4)
2/6.90	5)
2/2.90	6)
2/8.90	
2/1.50	
2/1.50	
2/1.50	
Item	Allow for protecting all decorated surfs.
	FLOORS
6.00	1) 100 mm wall strip flg in 100 mm wide nailed to horizontal dwt battens (40x25mm fastened to surf in this close nailed into conc at 200 m c/cs.
4.80	

CLAUSES V12 (1) & (2) – full description of materials used – and classification into groups in relation to surfaces to which paper is applied, i.e. walls, ceilings etc.

CLAUSE N13 (1) – lineal metres description to include sizes and method of fixing.

CLAUSE N1 (10) – cross sectional area of skirting exceeds 0.002m2 – mitres/angles etc measured as enumerated item.

2/6.00	10 mm th render and 3mm fin ct of pva a.b.d. to walls.
3.00	1) &
2/4.80	prep wall surfs to hang wallpaper w/ adhesive
3.00	5) paste to pva walls.
2/8.90	(Wallpaper - P.C. £4.50/roll.)
3.00	
2/1.50	_SKIRTINGS._
3.00	
2/6.00	25x100 mm sawn splayed skirting plugged to brick.
2/4.80	& k.p.s. v (3) on dets n. ex 150mm joint.
	Ditto
4/1	Angles to sawn skirting bead (c.s.a = .003m²).

CLAUSE T4 & T13 – finish treatment to screed in relation to type of floor finish to be quoted in description.

CLAUSE T14 – description to include method of fixing.

CLAUSE N33 – given as a separate item stating the area concerned.

CLAUSE T35 – ditto as last item.

4·80	40mm # Sc screed (1:4) laid level
4·40	2) rac bed, travelled smooth over
7·50	3) thermoplastic tiles.
6·90	&
6·90	4) 3mm # thermoplastic
2·90	5) tiles to BS 2592 laid in proprietary adhesive on S/D screed.
8·90	
1·50	
1·50	6) 40mm # Sc screed a.b.d. but in compartments n.e.k. 4·00m² on plan.
1·50	&
	7) 3mm # thermoplastic tiles a.b.d. in compartments n.e.k. 4·00m² on plan.
Item	Allow for protecting Hard strip flrg (6·00 & 24·80)
Item	Allow for protecting tile (by 1·80×2·40 7·50×6·90 6·90×2·90 6·90×1·50, 1·50×4·50)

The items normally included in a schedule for windows are: (a) type number, shape and size, (b) location, (c) overall size, (d) type of wall in which the window is fixed, (e) wall finishes, (f) lintel, (g) sill treatment, (h) ironmongery, (i) glazing.

Door schedules are similar, but include also: (a) door frame or lining details, (b) architrave and covermoulds, (c) threshold treatment, (d) decoration.

Note The protection clause is particularly important in making allowance for: (a) covering thresholds, sills etc. with light temporary battens; (b) protecting edges of doors, mullions etc. in glazed screens temporarily with adhesive tape.

The following schedule gives a general indication of the common items of work measured in relation to doors, windows and internal finishes, and at the same time gives the corresponding clauses of the *Standard Method of Measurement of Building Works*, their implications and the respective units of measurement.

Schedule of SMM requirements for doors and windows

Item No.	Description of work	SMM reference	Unit of measure
	Doors		
1	Doors.	**N19**. Measured as an enumerated item and classed according to type of door. In sliding/folding glazed screens each leaf is taken as a separate door.	No.
2	Decoration.	**V5(1)**. Painting of door, complete description of type and number of coats of paint, distinction between internal and external treatments. Measured as a superficial item, including painting top and side edges.	m²
3	Ironmongery.	**N32**. Description to include type, kind and quality of ironmongery, constituent parts, plus the backing material and method of fixing. Measured as an enumerated item.	No.
4	Door frames and linings.	**N20**. Frames and linings to doors grouped together and measured as a lineal item. Separate measure of each member of the frame, i.e. jambs, cills, heads etc.	m
5	Architraves.	**N13**. Measured as a lineal item with the description including the cross-sectional sizes, all labours to be included in the description. Mitres only enumerated on larger members: clause N1(10).	m No.
6	Decorations (frames, linings and architraves).	**V3(2)**. Where the girth of a combined surface exceeds 300 mm, measured as a superficial item. If less than 300 mm, girth taken as a lineal item, in 150 mm stages. *Note:* priming to back of frames and linings treated in a similar manner.	m²
	Metal windows		
7	Metal windows generally.	**Q3**. Measured as an enumerated item with a full description of the unit.	No.
8	Building in metal casements.	**G48**. Enumerated as separate items, description including overall sizes of the unit.	No.
9	Glass in doors and windows.	**U2 & 3**. Description to include type and thickness of glass and method of glazing. Classification of different types of glass. **U4**. Measured as a superficial item in pane sizes, as clause U4(1).	m²

Item No.	Description of work	SMM reference	Unit of measure
10	Bedding edges of glass.	**U12**. Measured as a lineal item, stating the type of material in which the glass is bedded.	m
11	Protection.	**Q10**. Given as an item.	Item
	Timber windows		
13	Timber windows generally.	**N21**. Measured as an enumerated item,	No.
14	Bedding of frame.	**G43(3)**. Bedding wooden frames measured as a lineal item, with description of the pointing required.	m
15	Fixing of frame.	**N31**. Fixing cramps to frame enumerated.	No.
16	Ironmongery.	**N32**. See previous reference, for doors.	
17	Glazing.	**U2 & 3**. See previous reference, for metal windows.	
18	Painting to metal or wooden window frames.	**V5**. Measured as a superficial item giving description of pane sizes, with separation of internal and external treatments.	m^2
19	Protection.	**N33**. Given as an item.	Item
	Adjustment of openings		
20	Brickwork and blockwork.	**G14 & 26**. See previous references in the section dealing with brickwork and blockwork	m^2
21	Plasterwork to walls.	**T4 & 11**. See previous references in the section dealing with internal finishes	m^2
22	Decoration to walls.	**V3, 4 & 12**. As item 21.	m^2
23	Lintels over door/window openings.	**F18**. Measured as an enumerated item.	No.
		G48(3). Cutting and pinning ends would only be applicable in alteration work (in new work, lintel built in as work proceeds).	No.
		G14 & 26. Adjustment of equivalent area of brickwork or blockwork as previous.	m^2
24	Skirting across door opening.	**N13**. See previous reference in item 5 of this schedule.	m
25	Flooring in door opening.	**T8 to T14**. See previous references in the section dealing with internal finishes	m^2

Sequence of measurement for glazed screen to clubhouse

1. Measurement of framing timbers – head, sill, stiles, mullions and transoms enumerated as a composite item.
2. Glazing beads.
3. Bedding edges of glass in plastic edging strip.
4. Glass – classification into different pane sizes.
5. Building in of frame into opening.
6. Priming back of frame before fixing.
7. Cleaning and priming glazing rebates.
8. Prepare and varnish mahogany frame – separate measure of internal and external treatments.
9. Deductions for previous over-measure of hollow wall, plasterwork and decorative finish in opening.
10. Precast concrete boot lintel over opening.

11. Rebated ends to lintel.
12. Deductions for over-measure of hollow wall, plasterwork and decorative finish to area replaced by lintel.
13. Plasterwork and decoration to soffit of lintel.
14. Closing cavity to sides of opening.
15. Plasterwork and decoration to reveals of opening.

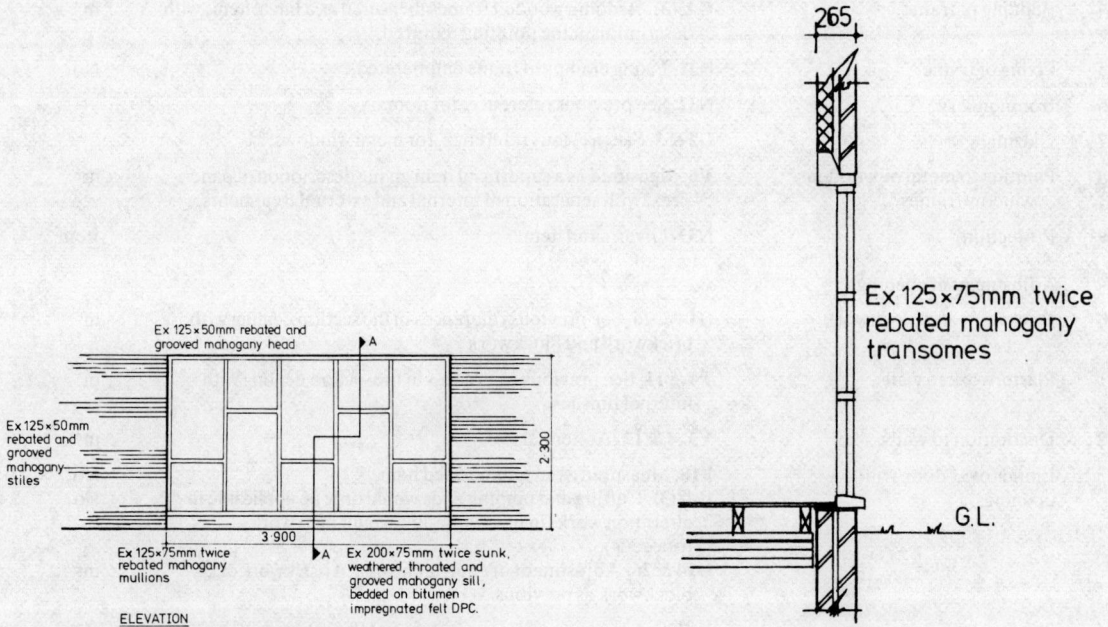

265

Ex 125 × 75mm twice rebated mahogany transomes

G.L.

SECTION A-A

Scale 1:50

Ex 125 × 50mm rebated and grooved mahogany head

A

Ex 125 × 50mm rebated and grooved mahogany stiles

2·300

3·900

A

Ex 125 × 75mm twice rebated mahogany mullions

Ex 200 × 75mm twice sunk, weathered, throated and grooved mahogany sill, bedded on bitumen impregnated felt DPC.

ELEVATION

All members of the frame to be properly jointed so as to ensure rigid construction.

All rebates to be well cleaned and primed prior to glazing.

Glass to be obscure patterned (Double flemish).

All glass to be bedded in plastic edging strip and fixed into frame with Ex 50 × 20mm mahogany glazing beads bedded in putty to BS 544 and fixed with brass cups and screws.

All surfaces of the frame to be well rubbed down and finished with two coats of polyurethane clear varnish.

Scale 1:100

GLAZED SCREEN – NORTH ELEVATION

CLUBHOUSE

Glazed ranking screen —
(o/a size 3·900 x 2·300 high)
comprising 1/ ex 125 x 50 mm
reb & grooved head;
ex 100 x 75 mm twice sunk
weath; twice grooved
sill, 1/ ex 125 x 50 mm
reb & grooved stiles,
2/ 125 x 75 mm twice
reb mullion, 1/ ex 125 x 75 mm
twice reb transomed
ex 50 x 20 mm glazing
beads bedd. putty to
B.S. 544 and fixed w
cups & screws finishing
all mitred

N.B./ Stiles mortice & tenoned
head & sill (haunched)
Mullins mortice & tenoned
head & sill (ordinary)
Transoms stub-tenoned
to mullions.

5/2 | 1.33
5/2 | 0.58
5/2 2 | 0.74

①

CLAUSE N22 – glazed screens, measured as enumerated item, with full description and if necessary a bill diagram should be provided.

BILL DIAGRAM.

2. DCE

3·900

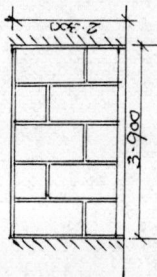

ELEVATION of SCREEN.

Bedding edges of glass in plastic edging strip

5/2 | 1.33
5/2 | 0.58
5/2 2 | 0.74

CLAUSE U12 – bedding edges of glass in particular edging materials measured as a lineal item, stating the material used.

Obscure double flemish glass.
(2/ 5 No panes
(av. 0·50-n ea
1·0 m²)

5/ 1.33
   0.74

Ditto in frames
(2/ 5 No panes
(av. 0·10 thick
0·50 m²)

5/ 0.58
   0.74

head & sill   2·300
                   ·150
                  ─────
                   2·150
2/3·900   7·800
5/2·150   4·300
                  ─────
                  12·100

12·10

Fixing frame in blg
bedd. mtl in c.m.
(1:8) on DPC mortar
bedd. frame bedd. in
c.m. (1:3) and pointed
one side in mastic
incldg all mitres

2/3/ 1

Galvd m.s. lug cramps
(40 x 3 x 250 mm lg)
o.e. bend & drilled
& screwed to back of frame,
other, o.e. fishtailed
& built into bed
joint of brick

②

CLAUSE U2 (1) – classification of work according to type and quality of glass.
CLAUSE U4 – glass grouped according to sizes of panes as U4(1).

CLAUSE G43(3) – building in frames including full description of all materials used, and measured as a lineal item.

CLAUSE N31 – enumeration of metal work fastenings to frame.

Sill 0·075
4/200 0·100
0·175

3·90

Ditto as last, but
ex 150→ nex 300mm
girth.
(Sill)
Letty
&

Ditto as last, but
ex 150→ nex 300mm
girth.
Batty

Leas 2·150
144·050
See·075 0·125
2·025

4/125 0·075
0·062½
0·137½

Ditto as last n-ex
150mm girth. (Mullions
+ Transoms)
Letty
&

4/ 2·03
5/ 0·71

Ditto as last n-ex
150mm girth
Batty

PAINTING

③

CLAUSE V1 (4) - measurement of work
required to be completed before fixing
frame. Painting less than 300 mm
wide - measured as a lineal item
with the girth of the work stated in
150 mm increments as CLAUSE V3 (2).

Priming back of frame
before fixg.
n-ex 150mm girth.

2/ 3·90
2/ 2·15

Colour + prime all
glazg. rebates. Prior
to fixg. glass.

5/2 1·33
5/2 0·74
5/2 0·58
5/2 0·74

Head 0·050
0·062½
0·112½
4/125

Stis 0·075
0·062½
0·137½

Rub down, clean
+ apply 2 cts of
polyurethane clear
varnish. (Head + sides)
n-ex 150mm girth.
Letty
&

2/ 2·15
3·90

Ditto as last
n-ex 150mm girth
Batty

ADJUSTMENT OF OPENING

NOTE – Deductions to be made for the previous over measure of the walling and finishes items.
Measurement procedure to follow previously stated rules.

3·90 2·15	Ddt HB skin coll wall in 65mm th blocks in cement (1:1:6) in cement bed, in gm (1:1:6) & fin of cement handling plg. & of coll wall
	Ddt 100mm th skin of thermalite tiles laid in gm (1:1:6) skin of coll wall
	Ddt formation of cavity 62½mm wide & coll walling incl 40×butterfly wall ties in th to BS1243.
	Ddt 10mm th render in cement backing + 3mm th fin ct of Carlite plr to BS1191. a blockwork
	Ddt proper 2 cts of emulsion pt to blk wall

Bearings 3·900   ·300
2/150   ·300
4·200

CLAUSE Q2 (1) – composite metalwork item fabricated in an offsite situation. Measured as an enumerated item and deemed to include for all work in lifting and fixing in position.
NB. With applied galVd, coating, no additional DPC bridging the cavity is necessary, hence no measurement of DPC at this point as would be the case with a precast concrete lintel.

BILL DIAGRAM

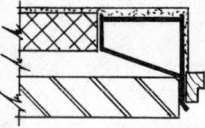

CNP4 – CATNIC LINTEL

NB. Inside face of CATNIC lintel covered in galvanised EML ready for application of direct plaster finish, therefore deduction of blockwork required.
NB. NO DEDUCTION OF FACING BRICKWORK.

CLAUSE T4 (2) – plasterwork not exceeding 300 mm wide, measured separately but still as a superficial item.
NB. Reveal – 120mm wide (SCALED)

1.	CATNIC pressed steel lintel type CNP4 (3 course lintel) hot dipped galv in compliance with BS729, clear opening 3.900m, clear snugly bearing 4.200m.
5·50 0·15	Ddt formation of cavity 62½mm wide a bd.
3·90 0·23	Ddt 100mm th thermalite skin of coll wall a bd.
3·90	10mm th render coat and 3mm th finish plr to BS1191 a bd to soffit of pressed steel lintel n.ex 300mm girth

CLAUSE V3 (2) - painting on general
surfaces - classification by situation.
NB. Not measured separately as an
isolated item as this work would be
completed at the same time as the
adjacent wall surfaces.

2/ 2·15	Bedding raw to sides of opg for glazed screen in Rubenoid Self-levod fitd polymer 102½ mm wide t/set verally in c·m (1:3)
2/ 2·15 0·12	10 mm th render 3mm th t·ct plga to BS 1191 a.bd to block th·i·f·t·l·l wall a.b.d. n·ex 300 mm·girth
2/ 2·15 0·12	Prep & 2 cts of emulsion pt to plga walls. (REVEALS).
3·90 0·12	Ditto to horiz sofft of pressed steel lintel (HEAD).
3·90 2·15	Rounded ext angle to plga work n·ex 10 mm radius & 7 mm fac jt f/plawork to mak'g discrements of pg.
Item	Allow for protecting all decorated surf.

⑦

7 Measurement of services installations

Covered by the rules of measurement laid down in the *Standard Method of Measurement of Building Works*, sections D, F, G, N, Q, T and V – limited number of references relating to excavation and builder's work; main referencs are sections R and W. Services installations cover various aspects:

- Plumbing installations.
- Drainage work.
- Electrical installations.

Plumbing installations can be further broken down into: (a) cold water installations, (b) hot water installations, (c) central heating installations etc. All work in each of these sections should be measured separately under the appropriate heading, although the general approach for the measurement of pipework and fittings will be the same in each case.

Plumbing installations

As a general guide the measurement should follow the sequence of installation, starting from the initial connection to the water undertaking's main (provisional sums being allowed to cover: (a) connection to mains and (b) reinstatement of the public highway) followed by the pipework, dealing with easily distinguishable sections of pipework runs followed by all the associated extra over items for fittings etc. plus any builder's work related to that pipe run.

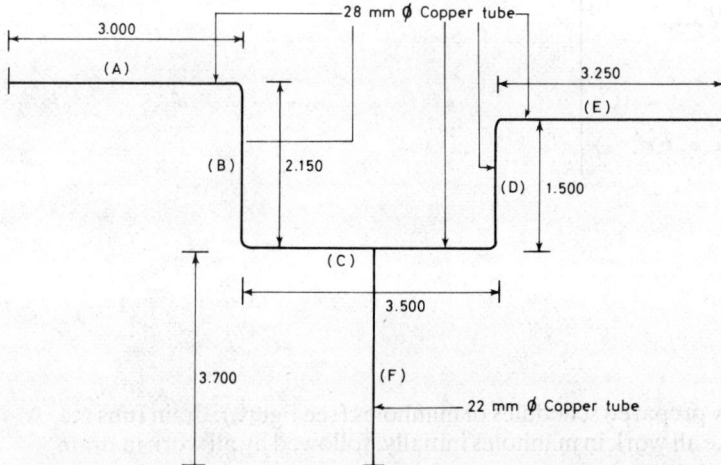

When measuring pipework the descriptions should classify the pipe according to the function it fulfills, i.e. flow, return, overflow, waste etc. In measuring the builder's work associated with pipe installations, pipes are classified in accordance with Clause G49: Small, n.ex. 55 mm diameter; Large, 55–110 mm diameter; Extra large, ex. 110 mm diameter

To ease the taking-off process and eliminate possible future errors with the measurement of complex installations, it is recommended that pipe runs should be coloured or shaded section by section as the measurement takes place. Pipework is measured over all fittings; thus in the diagram example, total length of 28 mm copper tube = A + B + C + D + E.

No deductions are made for the amount of pipework displaced by the fittings. Fittings are measured as extra over the pipework on which they occur; the builder's estimator is left to make the cost adjustment.

i.e. | Actual cost of fitting | = x pence |
| Less cost of pipework displaced | = y pence |
| ∴ E.o. cost of fitting | = $x - y$ pence |

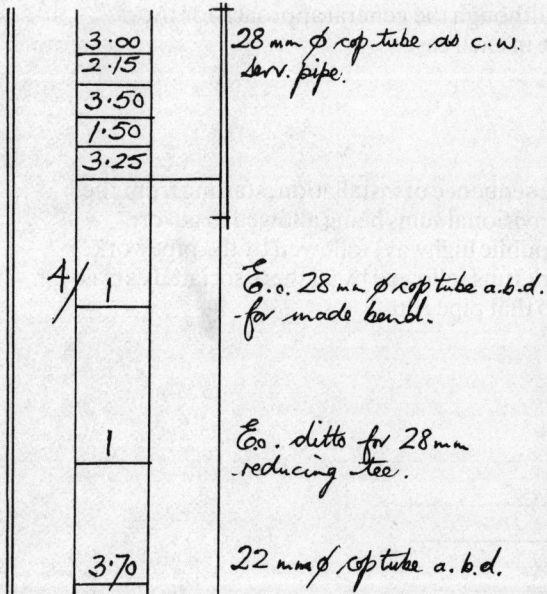

3·00	28 mm ⌀ cop tube do c.w.
2·15	serv. pipe.
3·50	
1·50	
3·25	
4/ 1	E.o. 28 mm ⌀ cop tube a.b.d. for made bend.
1	Eo. ditto for 28 mm reducing tee.
3·70	22 mm ⌀ cop tube a.b.d.

Below-ground drainage

Frequently measured from previously prepared schedules of manholes (see figure), drain runs etc. A suitable sequence of measure is to take all work in manholes initially, followed by all work in drain runs.

The measurement of the main drain runs starts at the head of the system working towards the sewer, at which point it will be necessary to allow provisional sums for: (a) connection to the public sewer, (b) reinstatement of the public highway. (In both cases the work is carried out by the appropriate public body.) The measurement of the branch drains then follows.

Following the measurement of manholes, drain runs etc., items must be allowed to cover the associated work of: (a) testing the system, (b) protecting the installation.

Manhole schedules simplify the presentation of the volume of information generally associated with complex drainage installations.

MANHOLE SCHEDULE

No.	Internal Size	Depth to Invert	Excavation Size	Straight Main Channel	Curved Main Channel	Branch Channel	Bends	Finish (Cover Slab)	Cover	Remarks

The approach to the measurement of excavation work in relation to drainage work must be carefully considered, because:

1. Excavation work, planking and strutting, backfilling and removal of excavated spoil associated with the construction of manholes are all measured as separate items, whereas
2. In the case of trench excavations for drain runs, all these items are grouped within a general description and the work measured as a linear rather than a cubic item.

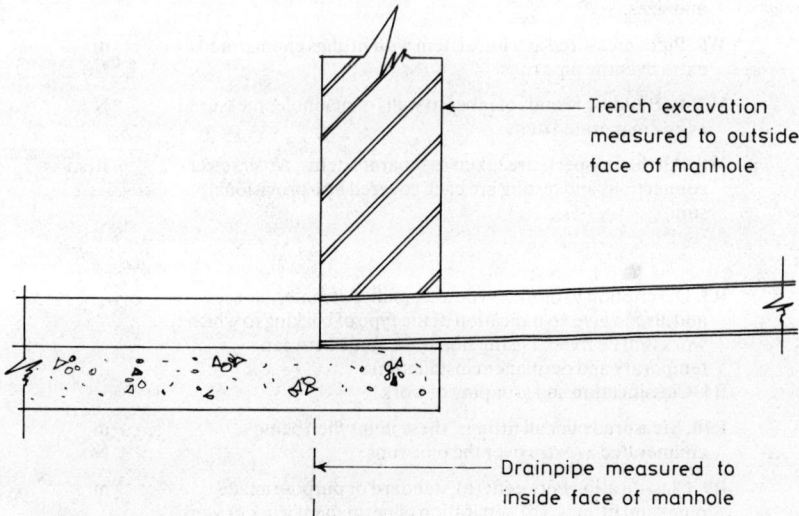

Trench excavation measured to outside face of manhole

Drainpipe measured to inside face of manhole

Electrical installations

This work is of a specialist nature usually undertaken on a sub-contract basis, the main contractor requiring his normal profit margin plus coverage of any attendance provided.

The initial entry of the power supply into the building is carried out by the regional electricity board, and from that point the rest of the installation is contractor's work. Electrical work covers: conduit, trunking and cable trays; cables; fittings and accessories; associated builder's work.

The following schedule gives a general indication of the common items of work measured in relation to services installations, and at the same time gives the corresponding clauses of the *Standard Method of Measurement of Building Works*, their implications and the respective units of measurement.

Schedule of SMM requirements for services installations

Item No.	Description of work	SMM reference	Unit of measure
	Drainage work below ground level		
1	Manhole construction generally.	Measurement of separate items of work in compliance with: D13 Excavation D14–24 Earthwork supports F4–5 Concrete work G5 Brickwork See previous references to these items in the section dealing with excavation and foundation work (page 00).	
2	Benching, channels, step irons and covers etc.	**W7(5)**. Measured as an enumerated item, description to include outline of materials, type, quality and overall sizes.	No.
3	Excavations for pipe runs.	**W3**. Measured as a lineal item complete, inclusive of excavation, earthwork supports, grading trench bottom, and any necessary backfilling, all these items to be fully described in the item.	m
4	Beds, haunching or encasing of drainpipes.	**W5**. Measured as a lineal item with full description of materials and sizes.	m
5	Pipes.	**W6**. Pipes measured as a lineal item with fittings enumerated as extra over the pipe runs.	m No.
6	Ends of pipes.	**W7(4)**. Building in ends of pipes to walls of manholes measured as an enumerated item.	No.
7	Sewer connections, testing and protection.	**W8**. All these aspects are taken as separate items. *Note:* sewer connections and testing are each covered by a provisional sum.	Item
	Cold and hot water installations		
8	Installations generally.	**R3**. Description to outline type and quality of materials used, and also to give an indication of the type of backing to which work is to be fixed. Distinction to be made between temporary and permanent installations. **R4**. Classification and grouping of work.	
9	Pipework generally.	**R10**. Measured over all fittings, these items then being enumerated as extra over the pipe runs.	m No.
10	Pipework in installations.	**R9**. Classification between: (a) standard or purpose-made pipes and fittings; (b) separation of permanent work or work to be temporarily fixed, dismantled and reassembled at a later date.	m
11	Joints on pipework.	**R10(1 & 2)**. Joints within the running lengths of pipework are included within the description of the pipes. Special joints measured as an enumerated item.	m No.
12	Labours and fittings on pipework.	**R13**. Labours such as made bends are measured as an enumerated item (as extra over). **R10(6)**. Fittings measured as an enumerated item.	No. No.

88

Item No.	Description of work	SMM reference	Unit of measure
13	Support to pipework.	R15. Type of support included in the description of the pipework.	m
		R40(5). Cutting and pinning ends of supports measured as a lineal item to pipes not exceeding 55 mm bore, and as an enumerated item when pipe bore exceeds 55 mm.	m No.
14	Sanitary appliances.	R22. Sanitary appliances measured separately as enumerated items, description to include ancillary items, i.e. chains and waste plugs, cantilever brackets etc.	No.
15	Valves etc.	R22. Valves and stop cocks etc measured as enumerated items and described as ancillaries.	No.
16	Water tanks and storage cylinders.	R22. Measured as enumerated items with full description of type, size and capacity etc.	No.
17	Insulation to services installations.	R30. Description to include type and quality of materials used.	
		R31. Insulation to pipework measured as a lineal item. R33(1). Insulation to tanks, boilers etc. generally measured as an enumerated item with sizes included in description.	m No.
	Associated builder's work		
18	Holes for pipework.	G49(1). Formation of holes for pipes through walls and G51, the subsequent making good after installation of pipework, are measured separately as enumerated items.	No.
19	Bedding and pointing.	R40(3). Bedding, fixing and pointing items such as sanitary fittings measured as an enumerated item.	No.
20	Cutting and pinning pipe supports.	R40(5). See previous reference to this clause in item 13 in this schedule.	m or No.
21	Making good around pipework.	Making good to plasterwork tiling etc., measured as an enumerated item in compliance with finishes section.	No.
22	Painting pipework etc.	Measured in compliance with SMM Section V.	
23	Testing of systems.	R37(5). Testing given as an item, stating method, duration and equipment to be used.	Item
24	Protection.	R41. Given as an item.	Item
	Electrical installations		
25	General aspects.	S3. Type and quality of materials used, method of fixing and type of background. S4. Classification and grouping of work. S5. Location of work as internal or external.	
26	Conduits.	S11. Measured over all conduit boxes and fittings etc. as a lineal item. Description of conduit to include relevant details of crampets etc. Special boxes enumerated as extra over items.	m No.
27	Electric cables.	S17(1 & 2). Cable runs measured as laid in lineal metres (net length of the conduit), description including details of type and size.	m

Sequence of measurement for cold water installation

1. Connection to water undertaking's main.
2. Underground service, plus fittings, bends and builder's work.
3. Rising main, vertical leg.
4. Fittings, bends and builder's work on item 3.
5. Rising main, through roof space.
6. Fittings, bends and insulation on item 5.
7. Branch to sink (kitchen), commencing with tee off rising main.
8. Low-pressure down service.
9. Fittings, bends and builder's work on item 8.
10. Branch pipe to bath, commencing with bend off down feed, bends and fittings up to and including tap.
11. Branch to lavatory basin, commencing with tee off down service.
12. Branch to WC flushing cistern, commencing with tee off down service.
13. Cold water storage tank and insulation.
14. Overflow pipe from tank and associated builder's work.

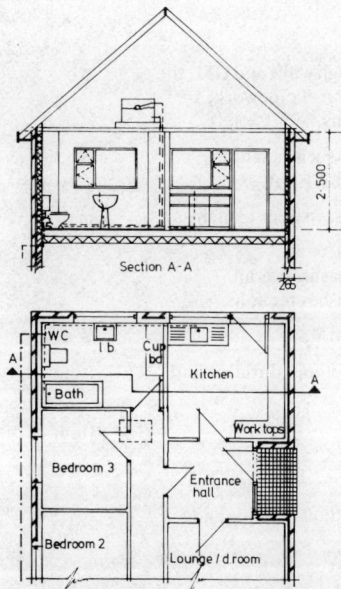

Section A-A

Scale 1:200

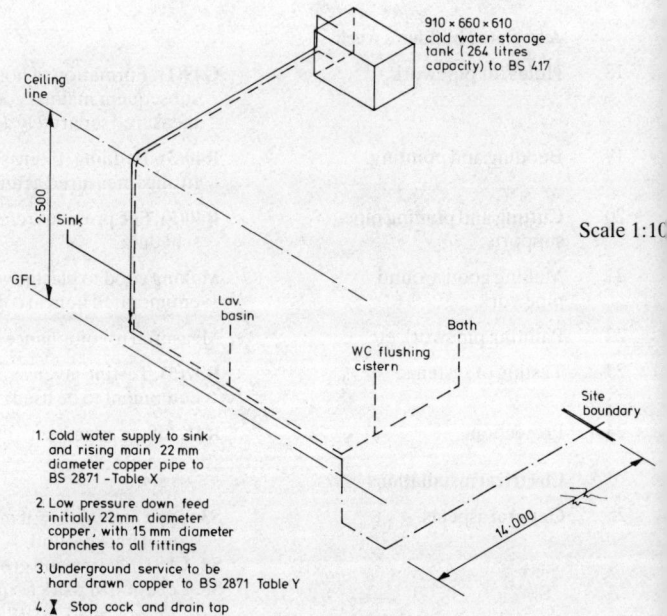

Scale 1:10

910 × 660 × 610 cold water storage tank (264 litres capacity) to BS 417

1. Cold water supply to sink and rising main 22 mm diameter copper pipe to BS 2871 - Table X

2. Low pressure down feed initially 22mm diameter copper, with 15 mm diameter branches to all fittings

3. Underground service to be hard drawn copper to BS 2871 Table Y

4. Stop cock and drain tap

SERVICE INSTALLATION

Conn. to Main.

Item

Allow prov. sum of £80 for conn. to water undertakings main in road of the existing 22mm ø cold pipe to the boundary incldg. to turns at stopcock and terminal enclosure.

&

Allow prov. sum of £30 for reinstatement of public highways.

CLAUSE R16 - allowance of provisional sum to cover this item of work to be completed by water undertaking.

YARD SERVICE
(up to icbbg stopcock)

4·000
4·350
14·350

CLAUSE B1 - separation of work carried out by (i) Public Bodies and (ii) Local Authorities

14·35

22mm ø hard drawn c.w. serv. pipe to BS 2871 table Y, incldg. conn. to BS 864 rillage joint etc.

&

CLAUSE R10 (1) - classified according to the function fulfilled by the pipe. Description to include size and type of pipe.

Excav. to provide level av. 500 mm dp. comm. at G.L. incldg. conslde. bttm. off, incldg. earthwork support, backfilling & conslde. fill in 200mm 111 compacted layers removal of surplus spoil.

CLAUSE D13 (8) - average depth of excavation for pipe trenches quoted to the nearest 250 mm unit.

①

E. o. 22mm ø hard drawn serv. pipe a. b. d. for made bend.

1.

pipe thru wall ·300
up to stopcock ·700
1·000

1·00

22mm ø hard drawn serv. pipe a. b. d. incldg. pipe clips to turns or pipe clips plugged & screwed to turns.

CLAUSE R13 - measurement of labours on pipes measured as enumerated items as "extra over" the pipe upon which they occur.

E. o. 22mm ø hard drawn serv. pipe a. b. d. for made bend.

1

Form hole thru' 265mm wall w. b. for s. p. & m. gd. surfs.

1

CLAUSE R15 - pipe clips included with description of pipes upon which they occur.

&

Ditto thru' 150mm w. inc. gd. floor surf. incl. bed. portion of d.p.c. for s.p. & m. gd.

&

C. & f. thermoplastic flds. to cut s. p.

1

22mm combined brass s/cock & drn. off tap incldg. conn. 2 No. to & db pipe incldg. coupling jt.

NOTE:- collection together of all builders work on this length of pipe up to stopcock.

②

CLAUSE R3(6) – classification of
background to which pipes are fixed.

CLAUSES R30 and 31 – insulation to
pipes – lineal item stating type
and bore.

CLAUSE R22 – classification of
ancilliary items, measured as
enumerated item with full description.

CLAUSE R10 (6) – fittings measured
"extra over" the pipework upon
which they occur.
NB. For "reducing tees" the largest
branch diameter only is quoted.

3·65	22 mm Ø cap c.w. slow a.b.d. + fixed w. clips but timber in roof space.
2/1	25 mm fibreglass wrap'g to 22 mm Ø cap pipe a.b.d, including binding in plastic coated jute straps.
1	E.o. 22 mm Ø cap c.w. slow pipe a.b.d. for made bend.
1	22 mm Ball valve to BS 1212, brass arm + plastic ball to BS 1968 + jt 22 mm cap pipe to custom union including straight conn.
	<u>SINK BRANCH</u>
1	E.o. 22 mm cap pipe a.b.d for 22 mm reducing tee.
	Horiz 1·000
	Conn ·150
	1·150
1·15	15 mm Ø cap c.w. slow pipe to BS 2871, including cap jts to BS 864, fix'd w. cap clips plugged + screwed to brickwork.

CLAUSES R10, R15 etc. a.b.d.

CLAUSE T11 – enumerated and description
to state classified size of pipe and
group as CLAUSE T11(2) according to
girth of opening.

NOTE: These lengths have been
scaled from the drawing.

	<u>HIGH PRESSURE RISING MAIN</u>
	Horiz 3·300
	Vert 2·500
	5·800
	Less bends above fl. level ·150
	5·650
	Tee incl'g jts ·100
5·55	5·550
	22 mm Ø cap c.w. slow to BS 2871, Table X, including cap jts to BS 864 fix'd w. cap clips plugged + screwed to brick at appt 1·350 m c/cs.
	Cut + fit 9·5 mm thk gypsum plaster bd to BS 1230.
	Mk. gd. 3 mm gypsum skim fin to BS 1191 and d.p.; n.ex 0·30 m girth.
2/1	E.o. 22 mm Ø cap c.w. slow pipe a.b.d. for made bend. (1 above GL + 1 thru' roof/space).
	Across 2·850
	half upstand ·800
	inlet ·800
	3·650

NB. taps are measured separately with the appropriate sanitary appliances.

CLAUSES R30 & 31 – insulation to pipes – lineal item stating type and bore.

Dim.	Description
1	jt 15 mm ∅ cap pipe to pillow tap (hot) necked taps to BS 1010 (SINK UNIT)
	LOW PRESSURE DOWN FEED.
3·05	To inlet ·1200 thro' roof 2·850 / 3·050 / 22 mm ∅ cap c.w. dern a.b.d. + fixd to timber in roof space a.b.d. (DOWN FEED)
	25 mm ∅ fibreglass wrapping to 22 mm ∅ cap pipe a.b.d.
2/1	E.o. 22 mm ∅ cap c.w. dern pipe a.b.d. for made bend.
1	jt 22 mm ∅ cap pipe to storage cistern inclg flanged tank conn'l + straight conn. Vertly 2·250 Horizly 3·250 Ditto 1·850 / 7·350

Dim.	Description
7·35	22 mm ∅ cap c.w. dern pipe a.b.d. + fixd to timber a.b.d.
2/1	E.o. 22 mm ∅ cap pipe a.b.d. for made bend.
	BRANCHES
	L. Basin
1	E.o. 22 mm ∅ cap dern a.b.d. for 22 mm reducing tee.
0·80	15 mm ∅ cap c.w. dern pipe a.b.d.
1	E.o. 15 mm ∅ cap dern pipe a.b.d. for made bend.
1	jt 15 mm ∅ cap pipe a.b.d. to pillow tap inclg 1No bent conn
	W/C
1	E.o. 22 mm ∅ cap dern a.b.d. for 22 mm reducing tee.

CLAUSE R10(6) – extra over item for reducing tee, stating largest branch diameter only.

2/1·50	ft 4/2x 3/400 1:200
	pgs 2/·50 :300
	1:500
1	Down duct bearers
	75 x 100 mm.
	(Tank Supports)
	Insulate to sides +
	top of tk a.b.l.
	Consist of lt.9 metal
	framing w. infill panels
	25 asbestos + 25 mm t
	of polystyrene, inclg
	removable cover.
	OVERFLOW
3·45	Horiz 2·900
	Vert 0·550
	3·450
	22 mm ø cap pipe a.b.d.
	as o/flow to tank
1	Form hole thro' 265 mm
	hollow wall + make good fair
	and s.p.
0·80	Prep, prime + (3)
0·80	to oxydised cap pipe
	hand.
	(VERT LESS)
	(to WK + LB)

CLAUSE N6 (3) – classified into appropriate grouping and measured as lineal item.

CLAUSE R33 – insulation to tank.

CLAUSE R10 (1) – classification according to function fulfilled by pipe.

length scaled from drawing.

NB. It is assumed that all other pipes in this exercise would be encased, this work being measured separately with also the separate measurement of associated painting, of the casing.

0·80	15 mm ø cap c.w serv
	pipe a.b.d. + fxg to
	BS.R a.b.d.
1	ft 15 mm ø cap c.w.
	serv pipe a.b.d. to
	union on finely
	cistern inclg JN°
	bent run.
	BATH
1	E.o. 22 mm ø c.w.
	serv pipe a.b.d. +
	made bend.
0·65	22 mm ø cap c.w.
	serv pipe a.b.d. +
	fxg to BS.R a.b.d.
1	ft 22 mm ø cap
	pipe to billow tap
	to BS.1010 on bath
	(m-/s)
	C.W. STORAGE TANK
1	C.W.S. Tank + cover.
	to BS 417 part 1, size
	910 x 660 x 610 mm dp
	264 litres cap affixed
	w. 2 N° conns for cap
	tube incl: parts for
	pipes in tank. Tank
	treated with 2 cts
	of bitumen paint.

CLAUSE R22 – full description stating the size, capacity and the number of connections provided on the tank.

Item	*Allow for all nec making good of holes etc for pipe runs in C.W. installation*
Item	*Allow for carrying out "air test" on completed C.W. installation.*
Item	*Allow for protecting all plumbing work in services installation.*

⑨

8 Measurement of staircase construction

Covered by rules of measurement laid down in the *Standard Method of Measurement of Building Works*, sections N, U and V. This work should be measured under a separate heading and can be further broken down into sections of:

- Concrete construction.
- Timber construction, including associated decoration work.

Concrete stairs

As with previous concrete items, work is dealt with in three sections: (a) supporting formwork, (b) reinforcement, and (c) concrete. The formwork item is the more involved, as the work must be classified and measured under the various categories listed in Clause F15.

The concrete and reinforcement items for the staircase construction complete are measured as one item for the work in the stairs plus the work in any associated landings. *Ref.* concrete clause F6(16), reinforcement clause F11(4f).

Balustrading to staircases can cover a wide range of items related to most sections of the *Standard Method of Measurement*, i.e. brickwork, concrete, timber panelling, metalwork and glazing.

Timber stairs

These are built up out of a wide range of members; these must all be fully detailed in the description of the staircase, which is measured as a composite unit and enumerated as Clause N23.

Dimension A + B = total girth of the step for the purpose of measuring painting to margins.

Note There will be one extra riser at the top of each flight.

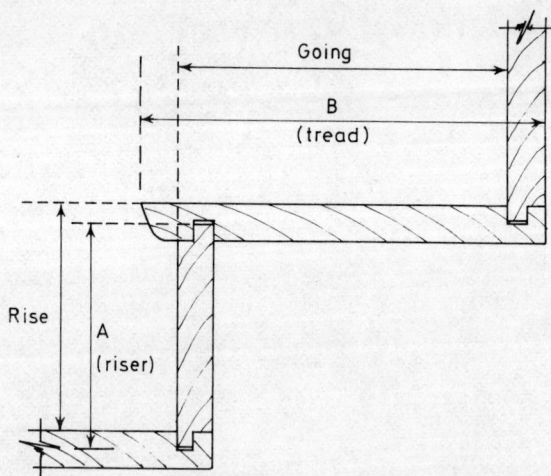

The trimming of openings for stairwells follows the procedure outlined for trimming openings in timber floors (Chapter 3), and generally will be measured with the floor.

The following schedule gives a general indication of the common items of work measured in the construction of staircases, and at the same time gives the corresponding clauses of the *Standard method of Measurement of Building Works*, their implications and the respective units of measurement.

Schedule of SMM requirements for staircase construction

Item No.	Description of work	SMM reference	Unit of measure
	Concrete stairs		
1	Concrete in staircase construction and steps.	Measured in compliance with F3 and F6(1), as dealt with in earlier references to concrete work.	
		F6(16). All concrete work relating to the construction of the flights of stairs should be grouped together and measured as a cubic item, i.e. including strings and associated landings.	m^3
		F6(11). Work in concrete balustrades of solid construction measured separately as a superficial item.	m^2
2	Finishes to concrete work.	**F8**. Measured as a superficial item as extra over the concrete work, description to include type and thickness of material used.	m^2
3	Reinforcement to staircases.	**F11(1 & 4f)**. Bar reinforcement in staircases including strings and associated landings.	m
4	Formwork to staircases.	**F15(1)**. Classification and grouping. Formwork to soffits of stairs and soffits of strings measured separately as a superficial item.	m^2
		F15(5 & 8). Formwork to risers and strings of staircases measured as a lineal item, the depth to be included in the description in one of the following categories: not exceeding 250 mm; 250–500 mm; exceeding 500 mm.	m
	Timber stairs		
5	Timber work generally (composite items).	**N17**. General rules for timber work must be complied with.	
6	Timber staircases.	**N23**. Measured as an enumerated item, description to give full supporting details of the members used in the construction of the staircase.	No.
7	Handrails.	**N23**. Measured as a lineal item, where not forming part of the complete staircase. If they form an integral part of the staircase and are fabricated off site as part of the composite unit, they are included in the measurement of the basic staircase.	m
			No.
8	Balusters and newel posts.	**Q5(6)**. Isolated (metal) balusters are measured as an enumerated item, with a complete description of the sizes etc.	No.
		N24. Newel posts measured as a lineal item, with newel cappings etc. enumerated.	m
			No.
9	Painting to strings.	**V3(2)**. Measured as a lineal item and given in 150 mm stages of girth.	m
10	Painting to margins.	**V4(1c)**. Ditto as last.	m
11	Prepare and varnish on general surfaces (panels, hand and middle rails).	**V3(2)**. Measured as superficial item (as over 300 mm girth).	m^2
12	Painting to metal balusters.	**V3(2)**. Measured as lineal item and given in 150 mm stages of girth.	m
13	Plasterboard treatment to underside of sloping soffit.	Measured as a superficial item in compliance with T17 and T5.	m^2
14	Emulsion paint treatment to plasterboard soffit.	Measured as a superficial item in compliance with V3 and 14(1c).	m^2
15	Protection.	**N33**. Given as an item.	Item

Sequence of measurement for timber staircase

1. Wrot softwood staircase (measured as a composite unit with full description).
2. Painting to strings and margins.
3. Newel posts and varnish treatment.
4. Handrail and middle rail, and varnish treatment.
5. Metal balusters and painting.
6. Plasterboard and skim finish to sloping soffit.
7. Emulsion paint finish to sloping soffit.

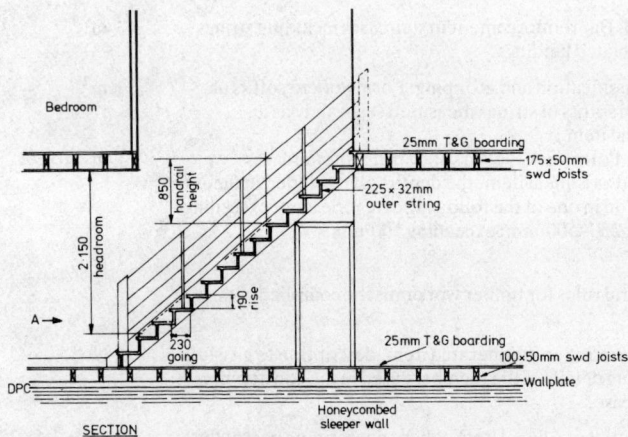

SECTION

Scale 1:100

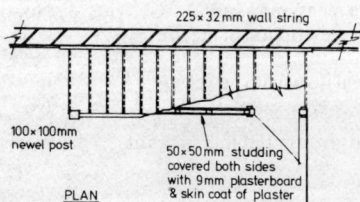

PLAN

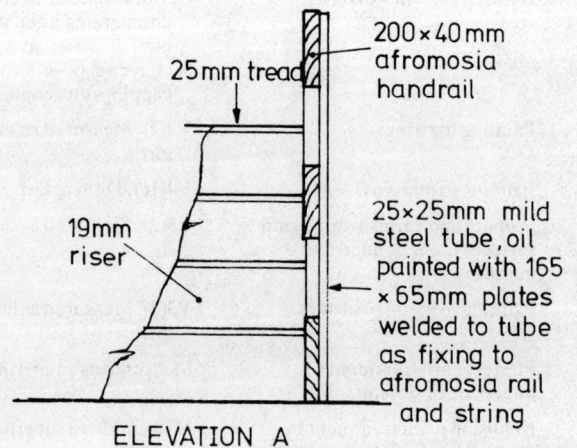

ELEVATION A

Scale 1:20

TIMBER STAIRS

The following work in one
number straight flight staircase
of 3.220 m going and 2.850 m wide.

CLAUSE N23 - composite unit fabricated in an "offsite" situation.
Full description and component details must be given.

1	Whot but and side light timber staircase as constructed/described to BS 585, consisting of 25 mm crossed/tongued treads in Sapeli moulded ×19 mm and/raised housed to treads at both ends or blocked together. Ends of steps housed one side ×32 ×225 mm moulded wall string plugged+screwed to wall and other side to 32×225 mm moulded outer string. The riser/plinth of both to supply timber and floor and out to top to suit moulded across the stairwell.
4.55	k.p.s ÷ ③ m string (wall string ex 150 ÷ n.e. 300 mm girth. (LENGTH SCALED)

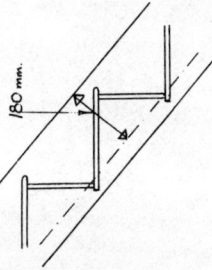

BILL DIAGRAM
180 mm.

CLAUSE V3 (1 & 2) - measured as a lineal item not exceeding 300mm girth.
NOTE - separation of inner and outer strings.

①

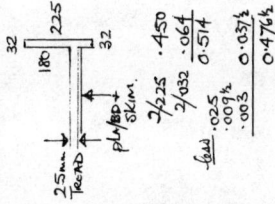

32   225
180
plumbed + 32
skim.
25 mm
tread

        2/225 ·450
        2/032 ·064
               0·514
less ·025 ·009½
      ·003   0·037½
            0·476½

CLAUSE V3(2) measured superficial, one continuous area over 300mm girth.

| 4·55 | | k.p.s ÷ ③ m str. ÷ onto face o/outer string |
| 0·48 | | |

PAINTING to MARGINS

14/2	0·44	less 0·860
2/	0·19	0·450
		÷2 0·410
		0·205
		head 0·230
		riser 0·190
		nos 0·020
		0·440

k.p.s × ③ to flr
ex 150 ÷ n.ex. 300
mm girth.

| 2/ | 1·35 | Ex. 100 × 100 mm Afromosia newel posts inc/leg bolting to first floor finishes wall below/ft in clay typ/chamfered cut to top to suit. (LENGTH SCALED). |

CLAUSE N24 - measured as a lineal item - stating method of fixing.

②

CLAUSE V3(2) - less than 30cm width, therefore measured as lineal item, stating girth in stages of 150mm.

CLAUSE T16 - measured as superficial item and grouped as CLAUSE T7(1) ie. all work relating to staircase construction is measured separately.

CLAUSE V4(1) work classified as to staircase areas.

3/1.00	Prep, prime + 3 coat metal to balusters n.e.x 150mm girth. *Itd*
4.10 0.86	9.5mm t gypsum pb/bd to BS 1230 fixed to soffit of stair.incl area. & 3mm t skim.ct of gypsum plaster to BS 1191, on pb/bd to soffit n.e.c. & prep + 2 cts of emulsion pt to pla clgs to staircase area.

NB/ ASSUME SPANDREL PANELLING TO UNDERSIDE OF STAIRCASE TO BE MEASURED LATER.

(4)

CLAUSE N13(1) - measured as lineal item, stating finish treatments and cross sectional sizes. Jointing to newel posts enumerated.

CLAUSE Q5 (6) - enumerated separately giving full description and sizes of members.

1.15 0.40	R.whbhp. 1:350 / 1:200 / 1:150. Prep, clean + 2 cts of mult.for.polyurethane varnish to hdrails. *Itd*
2/4.50	Ex 40x200mm of Australian hardwood/ handrail in chamfered, rounded to either backflap and newel balusters or ends housed to newels. (mould.dep) (LENGTH SCALED).
2/2/1	Housing 2 ends of rails to newel posts. 0.200 0.040 x2/0.240 0.480
2/4.50 0.48	Prep, clean + 2 cts of mult.for.polyurethane varnish to newel posts. *Itd*
3/1.	25x25mm. m.s.Tube baluster 1.0m long w 3No backplate size 165x65mm rebated on end fixg to dtl rails.

(3)

9 Measurement of external works

Covered by the rules of measurement laid down in the *Standard Method of Measurement of Building works*, sections D, F, Q, T and X. There are a wide range of aspects:

1. Paved areas such as car parks, school playgrounds etc. in in situ concrete, tarmacadam, precast concrete paving flags.
2. Paths, drives, roadways etc. in similar constructions to last.
3. Fencing and gates in a wide range of materials.
4. Landscaping work such as grassed areas, planting out (i.e. shrubs and trees).

The work should be measured and grouped separately under the above headings.

The following schedule gives a general indication of the common items of work measured in the construction of roads, paved and grassed areas etc., and at the same time gives the corresponding clauses of the *Standard Method of Measurement of Building Works*, their implications and the respective units of measurement.

Schedule of SMM requirements for external works

Item No.	Description of work	SMM reference	Unit of measure
	Paved areas, paths, drives and roadways etc.		
1	In situ concrete beds.	**F6(8)**. Measured as a cubic item in compliance with F5(2): not exceeding 100 mm; 100–150 mm; 150–300 mm; exceeding 300 mm. Concrete in beds forming roads, footpaths etc. described separately.	m³
2	Labours on concrete.	**F9(3 & 7)**. Finishes to faces – work on surface of unset concrete measured as a superficial item.	m²
3	Tarmacadam pavings.	**T4(1)**. Full description of type of materials, composition, mix and method of application. Measured in compliance with the rules relating to in situ floor finishes generally. **T8(1)**. Classification of slope (level, to falls or crossfalls etc.)	m²
4	Precast concrete paving work.	**T14(1)**. Description to include full details of materials and workmanship, measured in compliance with rules for in situ finishes (T4) as a superficial item. **T14(5)**. Cutting to angles, junctions etc. deemed to be included.	m²
5	Kerbs and path edgings.	**F19(1)**. Measured as a lineal item; intersections and angles etc. taken as enumerated items.	m No.
6	Beds and benchings to kerbs etc.	**W5**. Measured in lineal metres with full description of the mix and sizes.	m
	Fencing and gates		
	Fencing is dealt with in section X of the *Standard Method of Measurement* and is classified into: (i) open-type fencing, (ii) close-type fencing.		
7	Fencing generally.	**X2 to X12**. Classification into different types. Measured generally as a lineal item including struts, stays and posts etc., but special posts where applicable are enumerated separately in compliance with X13.	m No.

Item No.	Description of work	SMM reference	Unit of measure
8	Gates.	**X11**. Full description of the construction of the unit measured as an enumerated item. Gate stops enumerated separately.	No. No.
9	Holes for posts.	**X14**. Excavating holes for posts or forming mortices in concrete or brickwork etc. measured as enumerated items.	No.
	Landscaping		
10	Site preparation.	**D8 & 9**. Removal and preservation of turf and vegetable soil: see previous references to these aspects in the section dealing with foundation work (page 00).	m²
11	Removal of existing trees, hedges and undergrowth.	**D5, 6 & 7**. Removal of trees and roots (enumerated), hedges and roots (lineal), and bushes and undergrowth (superficial).	No. m², m
12	Seeding and turfing in landscaping work.	**D44**. Seeding grassed areas or turfing are measured separately as a superficial item.	m²
13	Planting new trees etc.	Not actually covered in the *Standard Method of Measurement*, but as a general guide the following approach can be adopted: (i) trees and shrubs enumerated; (ii) hedges measured as a lineal item.	No. m

Sequence of measurement for external works to clubhouse

1. *Tarmac road*. Including excavations, grading bottom, hardcore, base and wearing coats of tarmac. Precast concrete channels and kerbs, including concrete bed.
2. *Grassed areas*. Raking and preparing soil. Spreading and raking in seed, and rolling.
3. *Paths*. Including excavations, ash bed, mortar bed, precast concrete pavings. Edging kerbs and concrete bed.
4. *Fencing*. Fencing – full description of sections utilised. Extra over for gate posts. Gates and ironmongery.

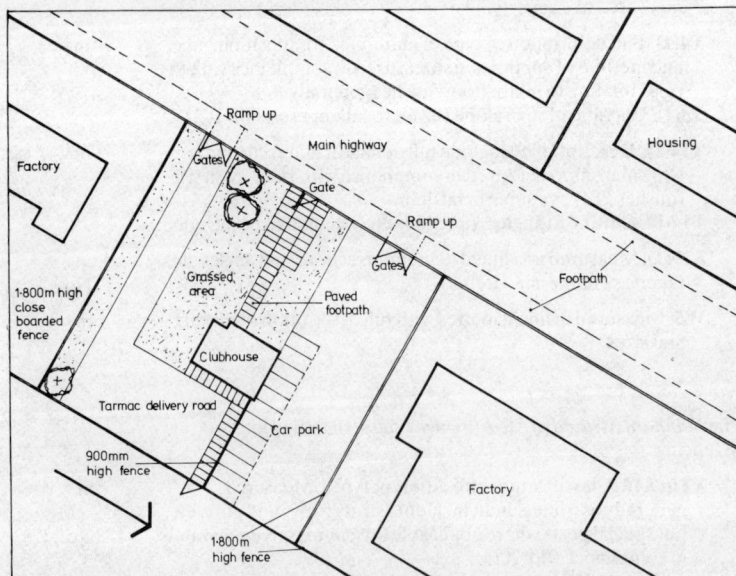

Scale 1:1000

Page ②

②

Handwritten column entries (dimension sheet):

Bed of hollow 150mm thick of granular matl blinded compacted & laid to falls to rec. Tarmac layer. (Method dep.)

29.50	
5.00	
20.80	
11.50	
27.50	
14.50	
14.70	
13.50	

Tarmac to BS 802 of 65mm thick base course of 40mm nom. size agg. and 25mm thick textured wearing course of 15mm nom. size agg. laid to falls crossfalls n.ex. 15° from line placed & compacted by some kind of roller. (Method dep.)

254 x 127mm p.c.c. channel to BS 340, bedded & jointed in c.m. (1:3).

2/	41.00
	15.50
	20.80
2/	41.00
	14.70
	0.50

Conc. bed (1:3:6/—20mm agg.) to underlined base to channel size 475 x 150mm dp.

152 x 305mm dp. splayed p.c.c. kerb, bedded & jointed in c.m. (1:3).

2/	41.00
	15.50
	20.80
2/	41.00
	14.70
	0.50

Typed notes:

CLAUSE D36 – measured as superficial item less than 250mm thick with the finished treatment of the layer to be included in the description.

CLAUSE T4(1) – full description of type of material and thickness of various layers.
CLAUSE T8 – classification in relation to finish, i.e. level or to falls or currents.

CLAUSE F19 – measured separately in lineal metres – stating sizes in description and method of laying. Ends, angles etc. enumerated.

BILL DIAGRAM

Page ①

①

CLUBHOUSE

EXTERNAL WORKS:

ROADS

NB. Dimensions to be scaled.

Excav. top soil av. 200mm dp.
(Dishg.) Rd(?)

29.50	
5.00	
20.80	
11.50	
27.50	
14.50	
14.70	
13.50	

Compact & regrade to falls & rec. sub-base of road & bed for channels. (Prelim. Ref.)

Hand excavate matl (topsoil) and level to 125mm thick in layers as surface of road to be graded later.

8/5/	29.50
	5.00
8/5/	20.80
	11.50
8/5/	27.50
	14.50
8/5/	14.70
	13.50

Typed notes:

CLAUSE D9 – top soil excavation superficial item stating average thickness to be removed.

NB. Excavt. 200mm thick, but matl. spread 125mm thick $\therefore$ to measure total area to be covered by vol of matl excavated a multiplying factor of $\dfrac{(200)}{(125)}$ must be applied. $= \dfrac{(8)}{(5)}$

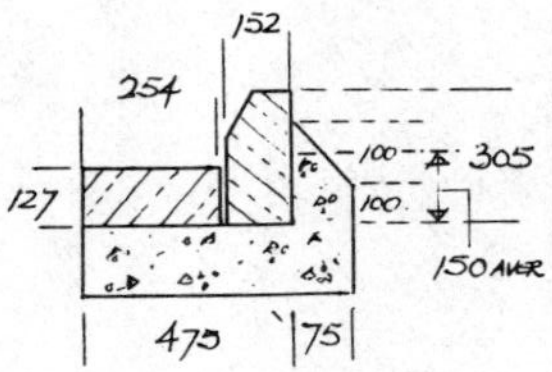

CLAUSE D40 – Preparation of ground measured initially with separate measure of seeding operation giving coverage of seed/m²

2/4

41.00	3.00	
18.50	20.00	
8.00	8.00	
2.50	1.50	

90° angles to p.c.c. kerb.

Rake & prep. soil incl. slight fertilizer in readiness for seeding.
&
Seeding with approved grass seed, including raking & rolling.
(Provide £0.48/kg/m²)

9.60	6.00
10.50	2.00

Ditto
Ditto as bst.

2/41.00	15.50
20.00	
2/41.00	
14.76	0.50

Conc. kerb to p.c.c. kerb 75mm wide by 150mm av. depth laid in contact with the grd.

PAVING

9.60	6.00
10.50	2.00
10.00	2.00
12.00	2.00

Excav. top soil av. 130mm dp, spread & levelled in 125mm bed & deposit on adjacent areas.
&
Compact grd & grade to falls a.b.d. for P.C.C. construction.

130/125	9.60
	6.00
130/125	10.50
	2.00
130/125	10.00
	2.00
130/125	12.00
	2.00

Spread excavated matl. (topsoil) & level in 125 mm bed on adj. areas to be grassed later.

2/9.60	6.00
2/2.00	
2/10.50	
2.00	
7.50	
12.00	

Excav. to the conc. bed to p.c.c. edging, n.e.x 0.30m wide and av. depth n.ex. 0.25m

NB. 130mm depth of soil excavated but spread and levelled in 125mm bed, therefore adjust by applying a multiplying constant of (130/125)

PATH EDGING

CLAUSE F19 - full description of sizes and method of laying.

2/ 9·60	51 x 203 mm o/p, p.c.c. flat edging w chamfered top edge to BS 340, bedd & jtd in c.m (1:3)
2/ 2·00	
2/ 10·50	
2·00	
7·50	Conc bed (1:3:6/- 20mm agg) and base to edging (detail) 150mm wide x 75mm dp.
12·00	Conc backg (1:3:6 as bd) 50mm wide by 113mm av. dept.

FENCING

CLAUSE X8 - measured as lineal item - but description to give full details and sizes of all members used in formation of fencing.

2/ 41·00	Oak cleft pale fencing to BS 1722 (part 5), 1800 - 41 off 2N horiz rails 75x75mm bored to 100x125mm posts) 2·500m bays Fix w 100x25mm th feather edged board ... fence heads at end Posts not 700mm into grd - in bag excavt and ... Conc. Bottom of panel fix w N°P.c.c. gravel bd 50x200th ... against short length of ... nailed the posts Top of panel fix w a 65x38mm cap'g twice weatherd on ... face nailed to ... rail
2/ 41·50	

(left sheet, page ⑤)

2/ 9·60	0·15	0·08
2/ 2·00	0·15	0·08
2/ 10·50	0·15	0·08
2·00	0·15	0·08
7·50	0·15	0·08
12·00	0·15	0·08

Remove exc April from site.

Deductions made to widths quoted in previous item on excav topsoil etc. for 50 mm thick edging slabs to either side of paths.

9·60	5·90
10·50	1·90
10·00	1·90
12·00	1·90

55 mm bed of grl hrdcd scatterd laid to falls to rec paving slabs & paving bed (mcavd) 20p

Screeded bed of mortar (1:1:6), 25mm b/t

P.c.c. paving slabs to BS 368, 50mm th size 600x750mm co-ordinated size laid on wet screeded bed of mortar jointed in fine matl

BILL DIAGRAM

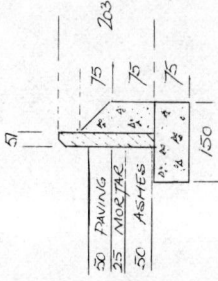

51, 203, 75, 75, 75, 150
50 PAVING, 25 MORTAR, 50 ASHES

⑤

| 13·50 | |

Ditto as last but 900mm hi. in 2No horiz rails a.b.d.+ 100×100mm posts; 1·500m long) [not into gd a.b.d.] + 25mm bolt fixed w. galv c.b.f. nuts + ditto do.

E.o. ditto bolt fencing for 175×175mm hi. 2·500m long gate posts w both into gd 1·600mm cp.

Excav holes for posts 600mm cp and casting in conc to support 150mm below surface + G.L.

NOTE – Four sets of gates.

CLAUSE X14(1) – excavating holes for larger posts to gates – measured as enumerated item.

| 4/2 | |

Close boiled gate size 2·000m by 1·800m hi. const of 25mm tt. bdg on 75×55mm framing and incldg g.i. drop catch hinges, adj socket + item 6 total sets.

CLAUSE X11 – as for fencing – full description but measured as an enumerated item.

| 6/1 | |

A checklist of books in the Butterworths Technician Series

BUTTERWORTHS **TEC** TECHNICIAN SERIES

MATHEMATICS FOR TECHNICIANS 1

FRANK TABBERER, Chichester College of Technology

This is an introduction to mathematics for the student technician, intended especially to cover mathematics at level one in TEC courses (core unit U75/005). The presentation will create an interest in the subject particularly for those students who have previously found maths a stumbling block. There are frequent examples and exercises, with a summary and revision exercise at the end of each chapter.

CONTENTS: Manipulating numbers. Calculations. Algebra. Graphs and mappings. Statistics. Geometry. Trigonometry.

192 pages   May 1978   0 408 00326 X

MATHEMATICS FOR TECHNICIANS 2

FRANK TABBERER, Chichester College of Technology

This covers mathematics at level two in TEC courses (units U75/012 and either U75/038 or U75 039), for those who have completed (or gained exemption from) the work in *Mathematics for Technicians 1*. It includes the alternative schemes of work allowed in the second stage of level two. The clear presentation and systems of examples and exercises, similar to those in the first volume, will enable students to gain a real grasp of the subject.

CONTENTS Trigonometry (1) Areas and volumes Statistics (1) Graphs Trigonometry (2). Equations and graphs. Mensuration Statistics (2) Introduction to calculus.

156 pages   September 1978   0 408 00371 5

PHYSICAL SCIENCE FOR TECHNICIANS 1

R. McMULLAN, Willesden College of Technology

This is intended for students studying the Physical Science level one unit of programmes leading to TEC certificates and diplomas. The text meets the requirements of the standard TEC syllabus for physical science (unit U75/004), a core unit of courses in building, civil engineering, electrical engineering and mechanical engineering. Attention has been paid to the visual presentation of the text, which is illustrated with diagrams and examples. Important concepts and formulae are clearly highlighted as an aid to learning and revision.

CONTENTS: Introduction. Fundamentals. Force and materials. Structure of matter. Work, energy, power. Heat. Waves. Electricity. Force and motion. Forces at rest. Pressure and fluids. Chemical reactions. Light.

96 pages   May 1978   0 408 00332 4

ELECTRICAL PRINCIPLES FOR TECHNICIANS 2

S. A. KNIGHT, Bedford College of Higher Education

Easy to read and in close conformity with the TEC syllabus, this book is intended primarily to cover TEC unit U75/019, Electrical Principles 2, an essential unit for both telecommunications and electronics students. The text includes examples, worked out for the reader, as well as problems for self-assessment, answers to which will be found at the end of the book. SI units are used exclusively throughout.

CONTENTS: Units and definitions. Series and parallel circuits. Electrical networks. Capacitors and capacitance. Capacitors in circuit. Magnetism and magnetisation. Electromagnetic induction. Alternating voltages and currents. Magnetic circuits. Reactance and impedance. Power and resonance. A.C. to D.C. conversion. Instruments and measurements. Alternating current measurements.

144 pages   May 1978   0 408 00325 1

ELECTRONICS FOR TECHNICIANS 2

S. A. KNIGHT, Bedford College of Higher Education

Provides an introduction to the basic theory and application of semiconductors. It covers the essential syllabus and requirements of TEC unit U76/010, Electronics 2, though some additional notes have been added for clarity. The text includes examples and self-assessment problems.

CONTENTS: Thermionic and semiconductor theory. Semiconductor and thermionic diodes. Applications of semiconductor diodes. The bipolar transistor. The transistor as amplifier. Oscillators. The cathode ray tube. Logic circuits. Electronic gate elements.

112 pages   June 1978   0 408 00324 3

BUILDING TECHNOLOGY 1 & 2

JACK BOWYER, Croydon College of Arts and Technology

These textbooks are primarily intended for the building technician taking TEC B2 construction courses. The clarity of text and illustrations should also, however, appeal to students of architecture and quantity surveying who need a good solid grounding in building construction.

BUILDING TECHNOLOGY 1

CONTENTS: The building industry. Site investigation, setting out and plant. Building elements: practice and materials. The substructure of building. The superstructure of building. Appendix: Building Standards (Scotland) Regulations 1971–75.

96 pages   March 1978   0 408 00298 0

BUILDING TECHNOLOGY 2

CONTENTS: First fixing joinery and windows. Services and drainage. Finishes and finishings. Second fixing joinery and doors. Site works, roads and pavings. Appendix: Building Standards (Scotland) Regulations 1971–75.

96 pages   May 1978   0 408 00299 9

HEATING AND HOT WATER SERVICES FOR TECHNICIANS

KEITH MOSS, City of Bath Technical College

By a system of nearly 200 worked examples, the author describes the routine design procedures for heating and hot water services in commercial and industrial buildings. Primarily intended for student HVAC technicians (TEC sector B3), it will also be useful for other students in sectors B2 and B3, and as a revision aid for experienced HVAC technicians encountering a change from Imperial to SI measurement.

CONTENTS: Heat energy transfer. Heat energy requirements of heated buildings. Heat energy losses from heated buildings. Space heating appliances. Heat energy emission. Heating and hot water service systems. The feed and expansion tank. Three-way control valves and boiler plant diagrams. Steam generation. Steam systems. Preliminary pipe sizing. Circuit balancing. Hydraulic resistance in pipes and fittings. Proportioning pipe emission. Hot and cold water supply. Circulating pumps. Steam and condense pipe sizing. Heat losses using environmental temperature. Medium and high pressure hot water heating.

168 pages   July 1978   0 408 00300 6